GW01157436

Lecture Notes
in Computational Science
and Engineering

122

Editors:

Timothy J. Barth
Michael Griebel
David E. Keyes
Risto M. Nieminen
Dirk Roose
Tamar Schlick

More information about this series at http://www.springer.com/series/3527

Alf Gerisch • Raimondo Penta • Jens Lang
Editors

Multiscale Models in Mechano and Tumor Biology

Modeling, Homogenization, and Applications

Springer

Editors
Alf Gerisch
Technische Universität Darmstadt
Department of Mathematics
Numerical Analysis and Scientific
Computing Group
Darmstadt, Germany

Raimondo Penta
School of Mathematics and Statistics
University of Glasgow
Glasgow, United Kingdom

Jens Lang
Technische Universität Darmstadt
Department of Mathematics
Numerical Analysis and Scientific
Computing Group
Darmstadt, Germany

ISSN 1439-7358 ISSN 2197-7100 (electronic)
Lecture Notes in Computational Science and Engineering
ISBN 978-3-319-73370-8 ISBN 978-3-319-73371-5 (eBook)
https://doi.org/10.1007/978-3-319-73371-5

Library of Congress Control Number: 2018933538

Mathematics Subject Classification (2010): 92-XX, 92C10, 92C15, 92-06

© Springer International Publishing AG, part of Springer Nature 2017
This work is subject to copyright. All rights are reserved by the Publisher, whether the whole or part of the material is concerned, specifically the rights of translation, reprinting, reuse of illustrations, recitation, broadcasting, reproduction on microfilms or in any other physical way, and transmission or information storage and retrieval, electronic adaptation, computer software, or by similar or dissimilar methodology now known or hereafter developed.
The use of general descriptive names, registered names, trademarks, service marks, etc. in this publication does not imply, even in the absence of a specific statement, that such names are exempt from the relevant protective laws and regulations and therefore free for general use.
The publisher, the authors and the editors are safe to assume that the advice and information in this book are believed to be true and accurate at the date of publication. Neither the publisher nor the authors or the editors give a warranty, express or implied, with respect to the material contained herein or for any errors or omissions that may have been made. The publisher remains neutral with regard to jurisdictional claims in published maps and institutional affiliations.

Printed on acid-free paper

This Springer imprint is published by the registered company Springer International Publishing AG part of Springer Nature.
The registered company address is: Gewerbestrasse 11, 6330 Cham, Switzerland

Preface

The International Workshop *Multiscale Models in Mechano and Tumor Biology: Modeling, Homogenization, and Applications* took place during September 28–30, 2015, at the Technische Universität Darmstadt, Germany. The workshop was organized by the research group *Numerical Analysis and Scientific Computing* of the *Department of Mathematics* at TU Darmstadt. The members of the Organizing Committee were Alf Gerisch, Jens Lang, and Raimondo Penta.

More than 35 researchers from different European universities and from various fields and application areas attended the workshop: mathematicians, engineers of different branches, mathematical modelers, biophysicists, computational scientists, and others. With this diverse background of participants one aim of the workshop was already achieved: bringing together an interdisciplinary group which can present, discuss, and combine theoretical as well as practical aspects of multiscale models as applied in the life sciences. Furthermore, research in mechano and tumor biology is growing together slowly but is performed largely by disjoint research groups. The workshop title and focus was chosen having this in mind and, for the mutual benefit of both research streams, we succeeded in bringing together members of both research areas.

The workshop was preceded by a four-hour course by Raimondo Penta introducing the concept of *Asymptotic Homogenization*, which forms the background of the introductory Chap. 1 of this book. This course was intended for the local graduate students and also for workshop participants not familiar with this technique. During the workshop we had 16 invited and contributed oral presentations as well as poster discussions during two dedicated poster session. The latter also greatly facilitated discussions and the informal exchange between participants.

In the aftermath of the workshop and inspired by the high quality of presentations and posters, we decided to publish this book containing the proceedings of the workshop. The chapters in this proceeding explore the broad spectrum of approaches in the current research on multiscale modeling in the life sciences in general and in mechano and tumor biology in particular: continuous vs. discrete vs. individual-based models, deterministic vs. stochastic models, spatially aggregated vs. spatially distributed models—their unifying theme being that they bridge scales in space and/or time. These models are investigated for their biological descriptiveness, their mathematical well-posedness, their computational accessibility, and last but not least for their suitability to advance the applied sciences to address urgent questions of biological understanding, medical progress, and biomimetic developments.

In the following, we give a brief overview of the chapters in this book, which we have arranged in three groups.

The first group of chapters, Chaps. 1 and 2, deals with the asymptotic homogenization technique.

Chapter 1, *An Introduction to Asymptotic Homogenization* by R. Penta and A. Gerisch, is different from all subsequent chapters in that it presents, in a tutorial style, the background and fundamental concepts of this approach to bridging scales. Asymptotic homogenization is a powerful mathematical tool used in multiscale modeling and it is being applied increasingly frequently, and at considerable complexity, in many applied multiscale settings. This chapter is intended for readers not yet sufficiently familiar with this topic. This way we aim at raising the interest of the audience in the technique and improve its chances to experience a further surge in applications.

This chapter is followed by the contribution of M.P. Dalwadi entitled *Asymptotic Homogenization with a Macroscale Variation in the Microscale* in Chap. 2, where the topic of Chap. 1 is expanded upon. Frequently, asymptotic homogenization is applied with the requisite assumption of a periodic structure. This is done as it typically simplifies the mathematical effort and, potentially more important, the computational work in simulations. However, this assumption is not necessary in general, see Chap. 1, and it is not always applicable in nature. Dalwadi tackles this problem and considers the toy problem of a drug diffusing past tissue to which it can be adsorbed. He shows how the assumption of a strict periodic structure can be relaxed to allow variation over a large length scale and how this can be exploited in a computationally feasible approach. The method presented can be used to include a slow structural variation in more complicated problems, adding a useful tool to the belt of the multiscale modeler.

The second group of chapters, Chaps. 3 to 5, considers various multiscale aspects of tumor biology. In Chaps. 3 and 4 the respective authors investigate different multiscale models of tumor/cancer cell invasion into the extracellular matrix, whereas Chap. 5 is concerned with the rigorous derivation of a macroscopic representation of fibrous tissue, like the extracellular matrix, from a microscopic individual-based description itself.

The contribution by S.A. Hiremath and C. Surulescu, entitled *Mathematical Models for Acid-Mediated Tumor Invasion: From Deterministic to Stochastic Approaches*, in Chap. 3 draws attention toward the growing evidence for the connection between tissue acidity and tumor invasion, and highlights the multiscale nature of tumor progression. The authors discuss some of the state-of-the-art deterministic models describing the effects of acidity on tumor invasion. This is immediately followed by a brief discussion about the inherent stochasticity in cancer dynamics and the presentation of two novel stochastic multiscale models featuring such random effects. Numerical simulations allow to shed light on some of the hidden dynamics of acid-mediated cancer migration that can rarely be observed in the framework of deterministic models, thereby confirming the importance of incorporating stochasticity in multiscale settings.

Chapter 4, *Numerical Simulation of a Contractivity Based Multiscale Cancer Invasion Model* by N. Kolbe, M. Lukáčová-Medvid'ová, N. Sfakianakis, and B. Wiebe, numerically investigates a particular multiscale model that describes the invasion of the extracellular matrix by cancer cells. Characteristic challenges of the model, which the authors address and overcome in their numerical approach, are the nonconstant advection and diffusion coefficients, the delay terms relating the different time scales relevant in the problem, as well as stiff reaction terms.

In the final chapter in this group, Chap. 5 entitled *Modelling Tissue Self-Organization: From Micro to Macro Models*, the authors P. Degond and D. Peurichard have in mind the example of tumor growth and the need to develop mathematical models which account for interactions at different scales. They focus on complex interconnected fiber networks, like the extracellular matrix in tumor or adipose tissue. Due to their simplicity and flexibility, the most frequently used models for such networks in the literature are individual-based models, which can incorporate any number of individual-level mechanisms. However, because these models are computationally challenging, continuous models are often preferred to study the biological systems at the macroscopic scale. In order to overcome the loss of information on the interactions at the individual level, a possible route is to derive a macroscopic model from a microscopic description. In this Chap. 5, the authors aim at providing a link between microscopic and macroscopic formulations, which account for mechanical interactions in biological tissues.

The third and final group of chapters, Chaps. 6 to 8, focuses on mechanical interactions (we could have included Chap. 5 here as well).

Chapter 6, entitled *A Multiscale Modeling Approach to Transport of Nano-Constructs in Biological Tissues* by D. Ambrosi, P. Ciarletta, E. Danesi, C. de Falco, M. Taffetani, and P. Zunino, considers the modeling of (functionalized) nanoparticles and their penetration into the living tissue through the vascular/capillary network. Any successful model in this field has to be multiscale and the authors consider geometrical as well as chemo-mechanical factors of the nanoparticle-tissue interaction. The central computational approach to bridge scales in their model is the *embedded multiscale method*. In this scheme, the capillaries are represented as one-dimensional channels embedded and exchanging mass in a porous medium.

G. Giantesio and A. Musesti investigate *A Continuum Model of Skeletal Muscle Tissue with Loss of Activation* in Chap. 7. Skeletal muscles have a highly ordered hierarchical structure; the elements at the microscale, such as the sarcomeres, have a very important role, for instance in the active behavior of the tissue. Although the model proposed in this chapter describes the behavior of the muscle from a macroscopic point of view, the microstructure enters in the form of the constitutive equations and in the choice of the parameters. The scales in the model are not yet coupled rigorously. Therefore it would be desirable to improve this model by directly starting from the microstructure of the tissue and using some homogenization technique to arrive at a macroscale description. The present chapter provides important groundwork for this ambitious future goal.

Finally, in Chap. 8, M. Zoppello, A. DeSimone, F. Alouges, L. Giraldi, and P. Martinon consider the *Optimal Control of Slender Microswimmers*. Within the framework of multiscale models they present a biomechanical model for a slender microscopic organism, swimming in water. The microscale modelling of the swimmer is crucial for the agreement of the model with reality and thus for the applications to future artificial devices.

Acknowledgments The members of the Organizing Committee thank Elke Dehnert, Sigrid Hartmann, Ursula Röder as well as the members of the research group *Numerical Analysis and Scientific Computing* at TU Darmstadt for their committed assistance in preparing the workshop. Furthermore, the generous financial support by the Graduate School of Excellence *Computational Engineering* at TU Darmstadt and by the DFG Priority Program SPP 1420 *Biomimetic Materials Research: Functionality by Hierarchical Structuring of Materials* is gratefully acknowledged. We are indebted to the Technische Universität Darmstadt for making available various university facilities throughout the workshop days. The publication of this book would not have been possible without the support of many individuals: first and foremost, the participants and speakers at the workshop and the contributors of chapters. We also thank Ruth Allewelt and the team at Springer-Verlag for their encouragement and professional support.

Darmstadt, Germany	Alf Gerisch
Glasgow, UK	Raimondo Penta
Darmstadt, Germany	Jens Lang
October 2017	

Contents

1 An Introduction to Asymptotic Homogenization 1
Raimondo Penta and Alf Gerisch
   1.1   Introduction 1
   1.2   One Dimensional Diffusion Problem 3
          1.2.1   Basic Set of Assumptions 3
          1.2.2   The Homogenized Problem 7
   1.3   Multidimensional Diffusion Problem 12
   1.4   Porous Media Flow: Homogenization of the Stokes' Problem 19
          1.4.1   Non-Dimensionalisation.. 20
          1.4.2   The Homogenized Problem 21
   1.5   Concluding Remarks.. 24
   References 25

2 Asymptotic Homogenization with a Macroscale Variation in the Microscale 27
Mohit P. Dalwadi
   2.1   Introduction 27
          2.1.1   Literature Review 28
          2.1.2   Chapter Outline 30
   2.2   Model Set-Up 30
   2.3   Homogenization.. 32
          2.3.1   Transforming the Normal 33
          2.3.2   Homogenization Procedure 34
   2.4   Interpreting the Homogenized Problem.............................. 38
   Appendix 1 39
   Appendix 2 41
   References 42

3 Mathematical Models for Acid-Mediated Tumor Invasion: From Deterministic to Stochastic Approaches 45
Sandesh Athni Hiremath and Christina Surulescu
   3.1   Introduction 45

	3.2	Continuum Mathematical Models: A Synopsis	47
		3.2.1 Deterministic Approaches	47
		3.2.2 Stochastic Approaches	50
	3.3	Multiscale Stochastic Models for Acid-Mediated Tumor Invasion	51
		3.3.1 Description of the Models	52
		3.3.2 Analytical Results	55
		3.3.3 Simulation Results	56
	3.4	Discussion	66
	Appendix		67
	References		68

4 Numerical Simulation of a Contractivity Based Multiscale Cancer Invasion Model 73
Niklas Kolbe, Mária Lukáčová-Medvid'ová, Nikolaos Sfakianakis, and Bettina Wiebe
    4.1 Introduction 73
    4.2 Mathematical Model 74
    4.3 Numerical Method 77
          4.3.1 Space Discretization 79
          4.3.2 Time Discretization 81
          4.3.3 Treatment of the Delay Term 81
          4.3.4 Choice of the Time Step 83
    4.4 Experimental Results 85
          4.4.1 Description of Experiments 87
    4.5 Conclusions 89
    References 89

5 Modelling Tissue Self-Organization: From Micro to Macro Models ... 93
Pierre Degond and Diane Peurichard
    5.1 Introduction 93
    5.2 Individual Based Model for Fibers Interacting Through Alignment Interactions 95
    5.3 Derivation of a Kinetic Model 98
    5.4 Scaling and Macroscopic Model 101
    5.5 Conclusion 106
    References 107

6 A Multiscale Modeling Approach to Transport of Nano-Constructs in Biological Tissues 109
Davide Ambrosi, Pasquale Ciarletta, Elena Danesi, Carlo de Falco, Matteo Taffetani, and Paolo Zunino
    6.1 Biophysics of Cancer 109
          6.1.1 An Overview of Transport Phenomena in Tumors 110
    6.2 A Microscale Approach to Transport of Nano-Constructs 116
          6.2.1 Microscopic Model 116
          6.2.2 Upscaling Method 120

Contents

		6.2.3 Numerical Methods	121
		6.2.4 Numerical Results	121
	6.3	A Macroscale Approach to Transport in Vascularized Tissues	126
		6.3.1 Governing Equations of Flow at the Macroscale	127
		6.3.2 Governing Equations of Mass Transport at the Macroscale	130
		6.3.3 Computational Solver	132
		6.3.4 Numerical Simulations	132
	6.4	Conclusions and Future Perspectives	134
	References		136

7 A Continuum Model of Skeletal Muscle Tissue with Loss of Activation 139
Giulia Giantesio and Alessandro Musesti
  7.1 Introduction 139
  7.2 Constitutive Model 141
    7.2.1 Passive Model 142
    7.2.2 Active Model 144
  7.3 Modelling the Activation on Experimental Data 148
    7.3.1 The Activation Parameter γ as a Function of the Elongation 149
    7.3.2 Loss of Activation 152
  7.4 Numerical Validation 152
  References 158

8 Optimal Control of Slender Microswimmers 161
Marta Zoppello, Antonio DeSimone, François Alouges,
Laetitia Giraldi, and Pierre Martinon
  8.1 Mathematical Setting of the Problem 161
    8.1.1 Kinematics of the N-Link Swimmer 162
    8.1.2 Equations of Motion 163
  8.2 Applications of the N-Link Swimmer 167
    8.2.1 Curvature Approximation 167
    8.2.2 N-Link Approximation of Sperm Cell Swimmer 168
  8.3 Controllability 171
    8.3.1 Classical Results in Geometric Control 171
    8.3.2 Main Theorem 173
  8.4 Minimum Time Optimal Control Problem for the N-Link Swimmer 176
    8.4.1 Minimum Time Problem 176
    8.4.2 Numerical Optimization 176
  8.5 Numerical Simulations for the Purcell's Swimmer ($N = 3$) 177
    8.5.1 The Classical Purcell Stroke 178
    8.5.2 Comparison of the Optimal Stroke and Purcell Stroke 179

8.6 Conclusions 181
References 181

Index 183

List of Contributors

François Alouges École Polytechnique CNRS, Palaiseau, France

Davide Ambrosi MOX, Dipartimento di Matematica, Politecnico di Milano, Milano, Italy

Pasquale Ciarletta MOX, Dipartimento di Matematica, Politecnico di Milano, Milano, Italy

Mohit P. Dalwadi Synthetic Biology Research Centre, University of Nottingham, Nottingham, UK

Elena Danesi MOX, Dipartimento di Matematica, Politecnico di Milano, Milano, Italy

Pierre Degond Imperial College London, London, UK

Antonio DeSimone Scuola Internazionale di Studi Superiori Avanzati (SISSA), Trieste, Italy

Carlo de Falco MOX, Dipartimento di Matematica, Politecnico di Milano, Milano, Italy

Alf Gerisch Technische Universität Darmstadt, Department of Mathematics, Numerical Analysis and Scientific Computing Group, Darmstadt, Germany

Giulia Giantesio Dipartimento di Matematica e Fisica, Università Cattolica del Sacro Cuore, Brescia, Italy

Laetitia Giraldi INRIA Sophia Antipolis Méditerranée, Team/équipe McTAO, Sophia Antipolis, France

Sandesh Athni Hiremath Felix-Klein-Center for Mathematics, Kaiserslautern, Germany

Niklas Kolbe Institute of Mathematics, Johannes Gutenberg-University, Mainz, Germany

Mária Lukáčová-Medvid'ová Institute of Mathematics, Johannes Gutenberg-University, Mainz, Germany

Pierre Martinon Team COMMANDS, INRIA Saclay - CMAP, École Polytechnique, Palaiseau, France

Alessandro Musesti Dipartimento di Matematica e Fisica, Università Cattolica del Sacro Cuore, Brescia, Italy

Raimondo Penta School of Mathematics and Statistics, University of Glasgow, Glasgow, UK

Diane Peurichard MAMBA—Modelling and Analysis for Medical and Biological Applications, LJLL—Laboratoire Jacques-Louis Lions, INRIA de Paris, Université Pierre et Marie Curie, Paris, France

Nikolaos Sfakianakis Institute of Applied Mathematics, Heidelberg University, Heidelberg, Germany

Christina Surulescu Felix-Klein-Center for Mathematics, Kaiserslautern, Germany

Matteo Taffetani Mathematical Institute, University of Oxford, Oxford, UK

Bettina Wiebe Institute of Mathematics, Johannes Gutenberg-University, Mainz, Germany

Marta Zoppello Universitá degli studi di Padova, Padova, Italy

Paolo Zunino MOX, Dipartimento di Matematica, Politecnico di Milano, Milano, Italy

Chapter 1
An Introduction to Asymptotic Homogenization

Raimondo Penta and Alf Gerisch

1.1 Introduction

Real world physical systems are usually *multiscale* in nature. They are characterized by strong heterogeneities, geometrical complexity, and different constituents which can interplay among several hierarchical levels of organization. Typical examples include, but are not limited to, fluid flow through geometrically complex and porous structure (encountered, for instance, when dealing with oil and gas recovery problems or physiological fluid flow through biological tissues and organs), as well as mechanical and chemical interactions among the various constituents of composite materials (such as, for example, soil or biological hard tissue, e.g. bone and tendons). From a modeling viewpoint, it is necessary to have a comprehensive understanding of the real world phenomena formulating qualitative and quantitative predictions (via analytical and numerical tools) to pursue validation against appropriate experimental data. As a matter of fact, this is basically a two-fold issue as (a) it is in general nontrivial and, especially for three-dimensional real problems, practically impossible to fully resolve *microscale* material and geometrical complexity and (b) experimental measurements usually provide average information on a *macroscale*, i.e. where the difference between different constituents cannot be easily detected. These arguments motivated the development of specific mathematical techniques

---

R. Penta (✉)
School of Mathematics and Statistics, University of Glasgow, G12 8QQ Glasgow, UK
e-mail: Raimondo.Penta@glasgow.ac.uk

A. Gerisch
Fachbereich Mathematik, AG Numerik und Wissenschaftliches Rechnen, Technische Universität Darmstadt, Dolivostr. 15, 64293 Darmstadt, Germany
e-mail: gerisch@mathematik.tu-darmstadt.de

designed to provide computationally feasible *macroscopic* mathematical models which, however, encode the crucial role of the *microstructure* (in terms of material heterogeneities, geometry, fine scale physical coupling, etc.). Although there exist various averaging techniques to obtain a macroscopic description of multiphase physical systems, such as the mixture theory (see, e.g., [4] and [5] for porous media), most of them lead to macroscale description of the system where information on the role of the microstructure is partially or entirely lost.

The *asymptotic homogenization* technique exploits the sharp length scale separation that exists in multiscale systems and a power series representation of the fields to provide macroscale systems of partial differential equations (PDEs) that satisfy both (a) and (b), as the derived models encode the role of the microstructure in their coefficients (hydraulic conductivities, diffusivities, elastic stiffness, etc.). As a drawback, the actual computation of the coefficients for multidimensional problems is only possible assuming appropriate regularity assumptions for the fields involved in the mathematical description (such as *local periodicity*). Furthermore, the microscopic description should be based on linearized balance equations (although nonlinear contributions can arise on a macroscale level) in order to decouple microscale and macroscale spatial variations of the fields.

Here, we introduce the technique via a very simple set of basic examples. We follow a direct approach widely explored in the literature (see, e.g. [2, 3, 13, 14, 16, 22]) which is well suited to introduce asymptotic homogenization to undergraduate/graduate students or scientists coming across this topic for the first time. Issues related to the theoretical foundation of the technique in terms of existence and uniqueness of the homogenized problems are not discussed here and we therefore refer the reader to the pioneering works [15][1] and [1] concerning *H-convergence* and *two-scale convergence*, respectively.

This book chapter is organized as follows:

- In Sect. 1.2, we start from the one-dimensional diffusion problem highlighting the concept of multiscale (spatial) variations and the basic assumptions that are needed to provide a macroscopic description of the problem via asymptotic homogenization. These include spatial variations decoupling, power series expansion, and local boundedness. We derive the diffusion-type homogenized problem and present the analytic form of the effective diffusion coefficient, which also holds for non-periodic microscale variations.
- In Sect. 1.3, we extend the one-dimensional formulation to the multi-dimensional diffusion problem and introduce the assumption of local periodicity, which is in this case necessary to compute the coefficients of the homogenized model. We show that the microscale information is encoded in the homogenized diffusion tensor, which can be computed solving a diffusion-type problem on a single periodic cell.

---

[1] An English translation can be found in [8], chapter 3.

- In Sect. 1.4, we present the asymptotic homogenization of the Stokes' problem, which leads to Darcy's law for porous media. In this case, the length scale separation is purely geometrical and is captured via an explicit non-dimensionalization process. The effective hydraulic conductivity is to be computed solving a Stokes'-type periodic cell problem.
- In Sect. 1.5, we present concluding remarks.

1.2 One Dimensional Diffusion Problem

We consider the one-dimensional diffusion-type boundary value problem (BVP)

$$\frac{d}{d\tilde{x}}\left(D(\tilde{x})\frac{du(\tilde{x})}{d\tilde{x}}\right) = f(\tilde{x}); \quad 0 < \tilde{x} < 1, \tag{1.1}$$

$$u(0) = a; \quad u(1) = b; \quad a, b \in \mathbb{R}, \tag{1.2}$$

where (1.2) represent non-homogeneous Dirichlet boundary conditions. Here, $f(\tilde{x})$ represents a known, spatially varying volume source, $D(\tilde{x})$ is the smooth, strictly positive, spatially varying diffusion coefficient, and $u(\tilde{x})$ the unknown scalar field. We assume that $f(\tilde{x})$ is regular enough such that a unique solution of (1.1–1.2) exists. BVPs of the type (1.1–1.2) are often encountered in the literature to model various physical phenomena, for example, the linear elastic displacement of an elastic rope, the temperature distribution for heat conduction, diffusion of pollutants, etc. Next we introduce the idea of multiscale spatial variations and formalize it via a basic set of assumptions.

1.2.1 Basic Set of Assumptions

We are interested in investigating the behavior of the problem solution $u(\tilde{x})$ when the diffusion coefficient $D(\tilde{x})$ exhibits *multiscale* spatial variations, i.e., when it displays a different behavior depending on the spatial resolution that we take into account. This is clearly highlighted in Fig. 1.1, where a representative solution of the problem (1.1–1.2) is plotted in the full domain (i.e. the unit length segment), and against a very small portion of it (zoomed in), where spatial variations on such a small scale can be clearly seen.

Next, we highlight the rigorous steps to deduce (a) the macroscopic profile of the solution of the one-dimensional diffusion problem and (b) how microscopic variations of the diffusion coefficient affect the macroscale behavior of the solution. At this stage, it is helpful to understand what *spatial scale separation* means in mathematical terms. Let us first introduce an informal, yet instructive argument. The problem (1.1–1.2) holds on the full unit segment, and our spatial coordinate $\tilde{x}$ spans

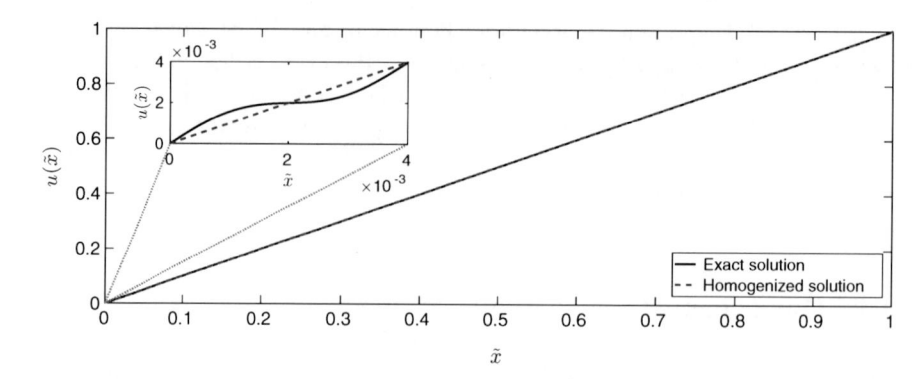

Fig. 1.1 The exact solution $u(\tilde{x}) = \frac{\tilde{x}+c\epsilon\,\sin(\tilde{x}/\epsilon)}{1+c\epsilon\,\sin(1/\epsilon)}$ of the BVP (1.1–1.2) for $a = 0$, $b = 1$, $D(\tilde{x}) = 1/(1 + c\cos(\tilde{x}/\epsilon))$, $\epsilon = 2/\pi \cdot 10^{-3}$, $c = 0.9$, $f = 0$ is plotted in black vs. the red-dashed homogenized solution $u(\tilde{x}) = \tilde{x}$ for $\tilde{x} \in (0, 1)$ (full plot) and $\tilde{x} \in (0, 4 \cdot 10^{-3})$ (inlay, zoomed in)

the full domain and represents the *physical* mapping for our problem. However, we aim at separating spatial macroscopic variations (see Fig. 1.1 in the domain $(0, 1)$), and microscale spatial variations that are detectable when "zooming in" (see Fig. 1.1 in $(0, 4 \cdot 10^{-3})$). To do this, we introduce a characteristic macroscopic length L (that is 1 for our specific problem), and another, much smaller characteristic microscopic length d (that is, for example, $4 \cdot 10^{-3}$ in the particular case shown in Fig. 1.1). We non-dimensionalize our physical spatial coordinate $\tilde{x}$ with respect to both the *microscale d* and the *macroscale L*, i.e.

$$\tilde{x} = Lx_M = dx_m. \tag{1.3}$$

Here x_M represents a non-dimensional coarse scale spatial mapping, as it is non-dimensionalized with respect to the macroscale L, whereas x_m represents a fine scale mapping, as it maps spatial variations resolved on the fine scale d. The two spatial coordinates are related by Eq. (1.3), i.e.

$$x_m = x_M/\epsilon, \tag{1.4}$$

where we define

$$\epsilon = \frac{d}{L}. \tag{1.5}$$

The small parameter ϵ measures the spatial scale separation between the microscale d and the macroscale L.

Remark 1.1 We remark that, even though a non-dimensional analysis is not always performed in the multiscale asymptotics literature, it is important to understand the relationship between macro and micro spatial variations of the fields. The

1 An Introduction to Asymptotic Homogenization

macroscale and microscale variables (also referred to as *slow* and *fast* scales, respectively) are typically denoted by x and $y = x/\epsilon$, respectively, and, using a commonly adopted abuse of notation, the physical variable is usually also denoted by x. The latter identification is rigorous when spatial scales decoupling is carried out after a non-dimensional analysis, as the physical spatial variable is already non-dimensionalized with respect to the macroscale L in that case. Such an analysis is highly recommended when dealing with multiphysics problems that typically comprise several parameters exhibiting different asymptotic behaviors with respect to ϵ, see, e.g. [19, 23]. However, the same, correct results are also obtained by avoiding an explicit non-dimensional analysis (as the local scale y will consistently stay dimensional and account for finer spatial variation the smaller ϵ is), provided that the correct asymptotic behavior of any variable and parameter involved in the differential problem is consistently taken into account. Here, we deal with a very simple problem (which is already in non-dimensional form, with macroscale length $L = 1$), so we just point out the nature of different spatial variables once and for all in this introductory section and avoid complicating the notation for the multidimensional problems illustrated in the following sections.

We are now ready to state the first crucial assumption

Assumption I (Length Scale Separation) *We assume that there exist two distinct spatial scales, referred to as the* microscale d *and the* macroscale L, *such that their ratio*

$$\epsilon = \frac{d}{L} \ll 1. \tag{1.6}$$

Our BVP (1.1–1.2) is currently stated in terms of the physical spatial scale $\tilde{x}$, which encodes both macroscale and microscale spatial variations. We aim to transform a single scale problem into a *multiscale* problem, and this leads us to the following assumption:

Assumption II (Spatial Variations Decoupling) *We assume that the unknown field u and the diffusion coefficient D that appear in the BVP (1.1–1.2) are functions of two formally independent spatial variables* $x = \tilde{x}$, *referred to as the* macroscale *and*

$$y = \frac{\tilde{x}}{\epsilon}, \tag{1.7}$$

(continued)

> **Assumption II** (continued)
> *referred to as the* microscale *variable. In particular, we may write*
> $$u = u(x, y), \quad D = D(x, y), \tag{1.8}$$
> *where*
> $$x \in (0, 1), \quad y \in (0, +\infty). \tag{1.9}$$

As a direct consequence of Assumption II, derivatives involving the physical spatial scale are now to be understood as *total (material)*, that is

$$\frac{d(\bullet)}{d\tilde{x}} = \frac{\partial(\bullet)}{\partial x} + \frac{dy}{d\tilde{x}} \frac{\partial(\bullet)}{\partial y} = \frac{\partial(\bullet)}{\partial x} + \frac{1}{\epsilon} \frac{\partial(\bullet)}{\partial y}. \tag{1.10}$$

We are interested in determining the macroscale behavior of differential problems of the type (1.1) in the presence of a sharp length scale separation between the macroscale and the microscale. Hence, it is convenient to consider a regular multiscale expansion for our unknown variable, as follows

> **Assumption III (Power Series Expansion)** *We assume that the multiscale unknown $u(x, y)$ can be formally represented by a regular expansion in power series of ϵ, i.e.*
> $$u(x, y) \equiv u^\epsilon(x, y) = \sum_{l=0}^{\infty} u^{(l)}(x, y) \epsilon^l. \tag{1.11}$$

The reader interested in rigorous issues related to the power series representation (1.11), which is appropriate under suitable regularity assumptions (even weaker than those assumed here), can refer to [10].

Accounting for Assumptions II and III, it seems that we have greatly complicated our problem (1.1), as we are now dealing with one more spatial variable and with infinitely many unknowns $u^{(l)}$. However, our aim is to determine the macroscale behavior of the problem solution whenever the length scale separation that characterizes the problem is sufficiently sharp, that is, for $\epsilon \to 0$. Thus, we will exploit our assumptions and the properties of the various coefficients to derive a macroscale differential problem for the leading order term of the multiscale power series expansion, i.e. $u^{(0)}$.

In order to prevent our multiscale functions from forming singularities with respect to the newly introduced microscale variable y, we also need a number of regularity requirements.

Assumption IV (Local Boundedness and Regularity) *We assume that*

- *Every field $u^{(l)}$, the external source f, and the coefficient D retain, with respect to the macroscale variable x, the same smoothness that characterizes the fields $u(\tilde{x})$, $f(\tilde{x})$, and $D(\tilde{x})$ appearing in (1.1) with respect to the variable $\tilde{x}$.*
- *Any function $u^{(l)}(x, y)$ that appears in (1.11) is locally bounded, i.e.*

$$\lim_{y \to +\infty} |u^{(l)}(x, y)| < +\infty \quad \forall x \in (0, 1) \text{ and } \forall l \in \mathbb{N}. \tag{1.12}$$

- *The multiscale diffusion coefficient $D(x, y)$ is strictly positive, locally bounded in the sense of (1.12), and there exist two strictly positive smooth functions $D_m(x)$ and $D_M(x)$ satisfying, for every $y \in (0, +\infty)$ and for every $x \in (0, 1)$*

$$D_m(x) \leq D(x, y) \leq D_M(x). \tag{1.13}$$

- *The volume source f is y-constant for the sake of simplicity, i.e.*

$$f = f(x). \tag{1.14}$$

We are now ready to apply the asymptotic homogenization technique to the problem (1.1). We intend to obtain a macroscale differential problem for the leading, zero-th order term $u^{(0)}$ that appears in the power series representation (1.11) of $u(x, y)$.

1.2.2 The Homogenized Problem

Let us enforce Assumptions I to IV. The multiscale problem associated to (1.1) then reads, by means of (1.10), as follows:

$$\epsilon^2 \frac{\partial}{\partial x}\left(D(x, y) \frac{\partial u^\epsilon}{\partial x}(x, y)\right) + \epsilon \frac{\partial}{\partial x}\left(D(x, y) \frac{\partial u^\epsilon}{\partial y}(x, y)\right) +$$
$$\epsilon \frac{\partial}{\partial y}\left(D(x, y) \frac{\partial u^\epsilon}{\partial x}(x, y)\right) + \frac{\partial}{\partial y}\left(D(x, y) \frac{\partial u^\epsilon}{\partial y}(x, y)\right) = \epsilon^2 f(x), \tag{1.15}$$

where we have multiplied both the right and the left hand side by ϵ^2 and u^ϵ denotes the power series representation (1.11). We then formally equate the same powers of ϵ in ascending order, starting from ϵ^0, in (1.15) until we obtain all the necessary conditions to derived a closed macroscale differential problem for the zero-th order component $u^{(0)}$. The derived *homogenized* problem will describe the one-dimensional diffusion process for well separated microscale and macroscale spatial variations, that is, for $\epsilon \to 0$.

ϵ^0

Equating the coefficients of power ϵ^0 in (1.15) yields

$$\frac{\partial}{\partial y}\left(D(x,y)\frac{\partial u^{(0)}}{\partial y}(x,y)\right) = 0. \tag{1.16}$$

We integrate over the microscale y and divide by $D(x,y)$ to obtain[2]

$$u^{(0)} = c_0(x)\int_0^y \frac{1}{D(x,s)}\,ds + c_1(x), \tag{1.17}$$

where c_0 and c_1 are y-constant functions to be determined. We now enforce Assumption IV (in particular relationship (1.13) concerning the existence of $D_M(x)$), to deduce

$$\int_0^y \frac{1}{D(x,s)}\,ds \geq \int_0^y \frac{1}{D_M(x)}\,ds = \frac{y}{D_M(x)}. \tag{1.18}$$

Hence, $u^{(0)}$ is not a bounded function of y unless $c_0(x) = 0$, that is

$$u^{(0)} = c_1(x) \tag{1.19}$$

only depends on the macroscale x. From now on, we thus simplify the notation and write $u^{(0)}(x)$.

ϵ^1

We equate the coefficients of power ϵ^1 in (1.15) and we account for the macroscale character of the leading order solution $u^{(0)}(x)$ to obtain:

$$\frac{\partial}{\partial y}\left(D(x,y)\frac{\partial u^{(0)}}{\partial x}(x)\right) + \frac{\partial}{\partial y}\left(D(x,y)\frac{\partial u^{(1)}}{\partial y}(x,y)\right) = 0. \tag{1.20}$$

---

[2]Note that the lower integration point is set to zero without loss of generality as the results are unchanged by integrating formally from any y_0 to y. Indeed, the homogenized coefficient that appears in (1.26) is invariant with respect to translations as $y - y_0$ is still approaching infinity as y approaches infinity.

1 An Introduction to Asymptotic Homogenization

We integrate over the microscale y to obtain:

$$D(x,y)\frac{\partial u^{(0)}}{\partial x}(x) + D(x,y)\frac{\partial u^{(1)}}{\partial y}(x,y) = b_0(x), \qquad (1.21)$$

where b_0 is a y-constant function. Dividing by $D(x,y)$ and further integrating over the microscale y yields

$$u^{(1)}(x,y) = b_0(x)\int_0^y \frac{1}{D(x,s)}\,ds - \frac{\partial u^{(0)}}{\partial x}y + b_1(x), \qquad (1.22)$$

where b_1 is another y-constant function. We have to ensure that $u^{(1)}$ stays locally bounded as $y \to +\infty$ (cf. Assumption IV). We first notice that, applying (1.13), the integral function $\int_0^y \frac{1}{D(x,s)}\,ds$ is bounded from below and above, i.e.

$$\frac{y}{D_M(x)} \leq \int_0^y \frac{1}{D(x,s)}\,ds \leq \frac{y}{D_m(x)}, \qquad (1.23)$$

and it becomes unbounded as y approaches $+\infty$. However, the term $\frac{\partial u^{(0)}}{\partial x}y$ is also unbounded as $y \to +\infty$ and both these contributions are $O(y)$. Therefore, we impose that they balance each other as y approaches $+\infty$, i.e.

$$\lim_{y\to +\infty}\left(b_0(x)\int_0^y \frac{1}{D(x,s)}\,ds - \frac{\partial u^{(0)}(x)}{\partial x}y\right) = 0, \qquad (1.24)$$

where the right hand side of (1.24) can be set to zero without loss of generality as any y-constant contribution could be previously taken into account redefining, for instance, the function b_1. The relationship (1.24) can be rewritten as

$$\frac{1}{\langle D^{-1}\rangle_\infty}\frac{\partial u^{(0)}(x)}{\partial x} = b_0(x), \qquad (1.25)$$

where we define

$$\langle D^{-1}\rangle_\infty := \lim_{y\to +\infty}\frac{1}{y}\int_0^y \frac{1}{D(x,s)}\,ds. \qquad (1.26)$$

We differentiate relationship (1.25) with respect to the macroscale x to obtain

$$\frac{\partial}{\partial x}\left(\frac{1}{\langle D^{-1}\rangle_\infty}\frac{\partial u^{(0)}(x)}{\partial x}\right) = \frac{\partial b_0(x)}{\partial x}. \qquad (1.27)$$

The differential problem (1.27) actually holds on the macroscale only and it describes the behavior of the leading order field $u^{(0)}$. Hence, this problem represents precisely our mathematical goal, provided that we are able to close it via a relationship for the right hand side $\dfrac{\partial b_0}{\partial x}$ in terms of known quantities. As we shall see below, we can obtain such a condition exploiting the local boundedness properties of the second order term $u^{(2)}$.

$\boxed{\epsilon^2}$

We now equate the coefficients of power ϵ^2 in (1.15) to obtain:

$$\frac{\partial}{\partial x}\left(D(x,y)\frac{\partial u^{(0)}(x)}{\partial x}\right) + \frac{\partial}{\partial x}\left(D(x,y)\frac{u^{(1)}(x,y)}{\partial y}\right) +$$
$$\frac{\partial}{\partial y}\left(D(x,y)\frac{\partial u^{(1)}(x,y)}{\partial x}\right) + \frac{\partial}{\partial y}\left(D(x,y)\frac{u^{(2)}(x,y)}{\partial y}\right) = f(x). \tag{1.28}$$

Since from (1.21) we have

$$\frac{\partial u^{(1)}(x,y)}{\partial y} = \frac{b_0(x)}{D(x,y)} - \frac{\partial u^{(0)}}{\partial x} \tag{1.29}$$

then

$$\frac{\partial}{\partial x}\left(D(x,y)\frac{\partial u^{(1)}(x,y)}{\partial y}\right) = \frac{\partial}{\partial x}\left(b_0(x) - D(x,y)\frac{\partial u^{(0)}(x)}{\partial x}\right). \tag{1.30}$$

We substitute (1.30) in (1.28) to obtain, rearranging terms:

$$\frac{\partial}{\partial y}\left(D(x,y)\frac{u^{(2)}(x,y)}{\partial y}\right) = f(x) - \frac{\partial b_0(x)}{\partial x} - \frac{\partial}{\partial y}\left(D(x,y)\frac{\partial u^{(1)}(x,y)}{\partial x}\right). \tag{1.31}$$

We integrate over the microscale y, divide by $D(x,y)$ and further integrate over y to finally obtain the following expression for $u^{(2)}(x,y)$:

$$u^{(2)}(x,y) = (f(x) - \frac{\partial b_0(x)}{\partial x})\int_0^y \frac{s}{D(x,s)}\,ds - \int_0^y \frac{\partial u^{(1)}(x,y)}{\partial x}\,ds$$
$$+ d_0(x)\int_0^y \frac{1}{D(x,s)}\,ds + d_1(x), \tag{1.32}$$

where $d_0(x), d_1(x)$ are y-constant functions. We now notice that, by Assumption IV, $D(x,y)$ and $u^{(1)}(x,y)$ are bounded functions of y. Hence, the first term of the right hand side of (1.32) is $O(y^2)$, while the second and third terms are $O(y)$ in the limit

1 An Introduction to Asymptotic Homogenization

$y \to +\infty$. In particular, it is necessary to require that the $O(y^2)$ (which cannot be balanced by other terms for $y \to +\infty$) identically vanishes, which implies[3]

$$f(x) = \frac{\partial b_0(x)}{\partial x}. \tag{1.33}$$

We exploit relationship (1.33) to close the differential problem (1.27), such that, accounting also for the appropriate boundary conditions dictated by our original problem (1.1–1.2), the *homogenized* BVP for the leading order coefficient $u^{(0)}(x)$ reads:

$$\frac{d}{dx}\left(\bar{D}(x)\frac{du^{(0)}(x)}{dx}\right) = f(x); \quad 0 < x < 1, \tag{1.34}$$

$$u^{(0)}(0) = a; \quad u^{(0)}(1) = b; \quad a, b \in \mathbb{R}, \tag{1.35}$$

where $\bar{D}(x)$ is the *homogenized* diffusion coefficient defined by

$$\bar{D}(x) = \langle D^{-1} \rangle_\infty^{-1}(x) = \left(\lim_{y \to +\infty} \frac{1}{y}\int_0^y \frac{1}{D(x,s)}\,ds\right)^{-1}. \tag{1.36}$$

The BVP (1.34–1.35) formally resembles the original one (1.1–1.2). However, microscale variations are now smoothed out and the homogenized coefficient $\bar{D}(x)$, as defined by (1.36), is the *harmonic* mean of the original $D(x, y)$ and not the simple arithmetic average defined by $D_{\text{AVG}}(x) = \lim_{y \to +\infty} \frac{1}{y}\int_0^y D(x,s)\,ds$. In fact, the harmonic average is more representative of the trend of the coefficient throughout the whole domain, i.e. high amplitude oscillations that are present in a subset which is much smaller than the whole domain do not greatly contribute to the final result. The convergence, for decreasing value of ϵ, of the exact solution of (1.1–1.2) to the homogenized one (obtained by solving the homogenized problem (1.34) using the appropriate coefficient $\bar{D}$) versus the "averaged solution" (obtained by solving the problem (1.34) using the arithmetic average D_{AVG}) is shown for a particular choice of boundary conditions in Fig. 1.2.

We would like to conclude this section remarking that the technique is applicable to nonperiodic local variations of the fields, as shown in Fig. 1.3. The reader can replicate the examples shown in Figs. 1.2 and 1.3 computing the homogenized solution analytically and compare it to the actual solution of the BVP (1.1–1.2). The latter is to be computed numerically for the nonperiodic example shown in Fig. 1.3.

---

[3] In order for $u^{(2)}(x, y)$ to be a bounded function of y, it is also necessary to require that the $O(y)$ terms compensate each other. However, as long as we focus on the leading order approximation $u^{(0)}$, it is sufficient to exploit local boundedness with respect to the $O(y^2)$ term only, as the latter involves the macroscale function $b_0(x)$ (see Eq. (1.27)) that can eventually close the macroscale problem for $u^{(0)}$.

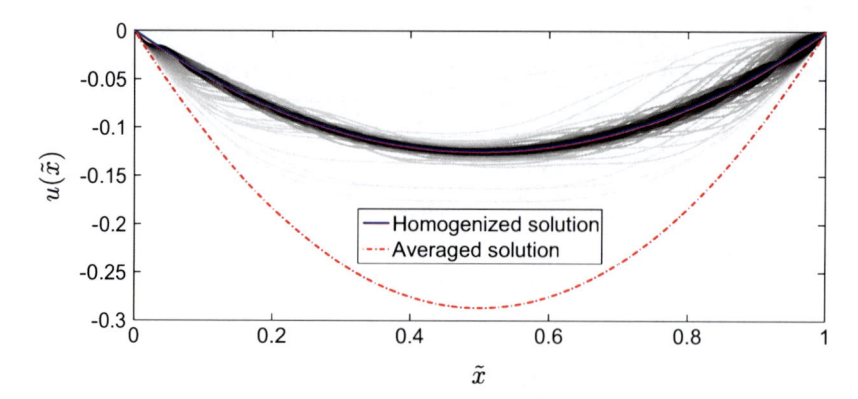

Fig. 1.2 The exact solution of the BVP (1.1–1.2) for $a = 0$, $b = 0$, $f = 1$, $D(\tilde{x}) = 1/(1 + c\cos(\tilde{x}/\epsilon))$, $c = 0.9$, and $0.01 < \epsilon < 1$, is shown in grey scale and it gets darker and darker the smaller ϵ becomes. The solution of the real problem is converging to the *homogenized* solution (shown in blue) and not to the averaged one, represented by the dash-dot line in red

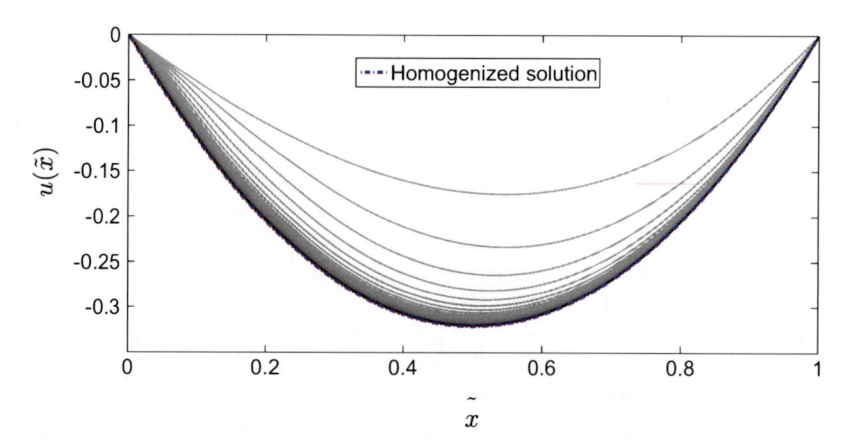

Fig. 1.3 The exact solution of the BVP (1.1–1.2) for $a = 0$, $b = 0$, $f = 1$, $D = (\arctan(x/\epsilon) + 1)^{-1}$, and $0.02 < \epsilon < 1$, is shown in grey scale and it gets darker and darker the smaller ϵ becomes. The solution of the real problem is converging to the *homogenized* solution represented by the dash-dot line in blue

1.3 Multidimensional Diffusion Problem

We aim to generalize the diffusion problem introduced in the previous section to dimension $n \in 1, 2, 3$. We then consider the following classical diffusion problem for a scalar field u in an open connected set $\Omega \subset R^n$ with smooth boundary $\partial \Omega$, i.e.

$$\nabla \cdot (D(\mathbf{x})\nabla u(\mathbf{x})) = f(\mathbf{x}), \ \mathbf{x} \in \Omega \tag{1.37}$$

1 An Introduction to Asymptotic Homogenization

equipped, for example, with non-homogeneous Dirichlet boundary conditions

$$u(\mathbf{x}) = g(\mathbf{x}), \ \mathbf{x} \in \partial\Omega. \tag{1.38}$$

Here, $D(\mathbf{x})$ is the strictly positive and smooth spatially varying diffusion coefficient, $f(\mathbf{x})$ a known volume source and $g(\mathbf{x})$ a known function dictating the behavior of the solution at the boundary. The functions f and g are assumed sufficiently regular such that a solution of the classical diffusion problem (1.37–1.38) exists. At this stage, we assume length scale separation, spatial variable decoupling (that implies transformation of differential operators) and power series representation, so that we can readily generalize assumptions (I–III), together with relationship (1.10). The multiscale, multidimensional problem associated to the n-dimensional diffusion problem (1.37) then formally reads

$$\epsilon^2 \nabla_\mathbf{x} \cdot (D(\mathbf{x},\mathbf{y})\nabla_\mathbf{x} u^\epsilon(\mathbf{x},\mathbf{y})) + \epsilon \nabla_\mathbf{x} \cdot (D(\mathbf{x},\mathbf{y})\nabla_\mathbf{y} u^\epsilon(\mathbf{x},\mathbf{y})) +$$
$$\epsilon \nabla_\mathbf{y} \cdot (D(\mathbf{x},\mathbf{y})\nabla_\mathbf{x} u^\epsilon(\mathbf{x},\mathbf{y})) + \nabla_\mathbf{y} \cdot (D(\mathbf{x},\mathbf{y})\nabla_\mathbf{y} u^\epsilon(\mathbf{x},\mathbf{y})) = \epsilon^2 f(\mathbf{x},\mathbf{y}), \tag{1.39}$$

where we have included microscale variation of the volume source for the sake of generality. Here, $\nabla_\mathbf{x}$ and $\nabla_\mathbf{y}$ represent the gradient with respect to the macroscale and microscale variables $\mathbf{x}$ and $\mathbf{y}$, respectively. We are using the same symbol $\mathbf{x}$ for the macroscale and the physical spatial variables for the sake of simplicity of notation. Since we introduced the additional microscale spatial variable $\mathbf{y}$, we then need to state appropriate regularity assumptions for every multiscale quantity that appears in (1.39). We could indeed generalize Assumption IV and apply the asymptotic homogenization technique to obtain a well-posed macroscale problem. However, whenever $n > 1$, we cannot in general obtain single closed form expressions for the homogenized coefficients (see, e.g., [13]). In general, the latter are to be computed solving microscale differential problems that in principle hold on the whole microscale domain (that extends up to infinity in the limit $\epsilon \to 0$). One of the most important goals for this asymptotic homogenization technique is to determine an *effective* differential problem that describes the macroscale behavior without resolving the full details of the microscale, thus enhancing computational feasibility. At the same time, the macroscale problem should retain information on the microscale encoded in the homogenized coefficient that should be readily accessible, as is the case for the integral (1.36) which defines the one-dimensional homogenized diffusion coefficient. As is widely enforced in the asymptotic homogenization literature, we can focus on a smaller portion of the microscale by assuming $\mathbf{y}$-*periodicity* of the fields involved in (1.39). We state the periodicity assumption in three dimensions, as it can be readily restricted for $n = 1, 2$.

> **Assumption V (Local Periodicity)** *There exists a family of vectors*
>
> $$\mathbf{R}(\eta, \kappa, \upsilon) := \eta \mathbf{I}_1 + \kappa \mathbf{I}_2 + \upsilon \mathbf{I}_3, \quad \eta, \kappa, \upsilon \in \mathbb{Z} \qquad (1.40)$$
>
> *with fixed vectors* $\mathbf{I}_1, \mathbf{I}_2, \mathbf{I}_3 \in \mathbb{R}^3$ *that constitute a basis of* $\mathbb{R}^3$, *such that, for every field that appears in* (1.39), *collectively denoted by* ψ, *we have*
>
> $$\psi(\mathbf{x}, \mathbf{y}) = \psi(\mathbf{x}, \mathbf{y} + \mathbf{R}(\eta, \kappa, \upsilon)), \quad \forall \eta, \kappa, \upsilon \in \mathbb{Z}. \qquad (1.41)$$

Note that Assumption V is stated for arbitrarily shaped periodic cells and that rectangular (cuboid in three dimensions) periodic cells are simply obtained assuming $\mathbf{I}_n \propto \mathbf{e}_n$ for every n. We therefore account for Assumption V (instead of generalizing local boundedness stated in (1.12)), appropriately extended to the external source $f(x, y)$, while we retain the remaining points of Assumption IV concerning the regularity of the diffusion coefficient and all the fields with respect to the macroscale variable $\mathbf{x}$. Thus, microscopic variations of multiscale fields can now be studied on a single periodic cell defined by the vectors $\mathbf{I}, .., \mathbf{I}_n$. A simple cartoon representing a two-dimensional rectangular cell is shown in Fig. 1.4.

We now proceed by equating the same powers of ϵ in ascending order, starting from ϵ^0, as we have done in the one dimensional case. It is important to bear in mind that we are assuming local periodicity and that the arising differential conditions, though retaining a parametric dependence in terms of the macroscale $\mathbf{x}$, hold on the periodic cell (which we also refer to as Ω to avoid complicating the notation) spanned by the microscale variable $\mathbf{y}$.

$\boxed{\epsilon^0}$

Equating the same powers of ϵ^0 in (1.39) yields

$$\nabla_\mathbf{y} \cdot \left(D(\mathbf{x}, \mathbf{y}) \nabla_\mathbf{y} u^{(0)}(\mathbf{x}, \mathbf{y})\right) = 0 \text{ in } \Omega. \qquad (1.42)$$

Fig. 1.4 A representation of a two-dimensional rectangular cell defined by the vectors $\mathbf{I}_1 = y_{p1}\mathbf{e}_1$ and $\mathbf{I}_2 = y_{p2}\mathbf{e}_2$, where $y_{p1}, y_{p2} \in \mathbb{R}^+$

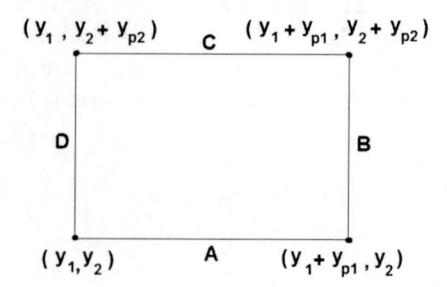

1 An Introduction to Asymptotic Homogenization

Relationship (1.42), equipped with periodicity conditions on $\partial\Omega$, constitutes a standard diffusion-type cell problem that admits a unique solution up to a **y**-constant function. In particular, any constant is also periodic and solves (1.42), thus we deduce that $u^{(0)}$ is independent of **y**, i.e.

$$u^{(0)} = u^{(0)}(\mathbf{x}). \tag{1.43}$$

$\boxed{\epsilon^1}$

We equate the same powers of ϵ^1 in (1.39) and account for (1.43) to obtain

$$\nabla_\mathbf{y} \cdot \left(D(\mathbf{x},\mathbf{y})\nabla_\mathbf{y} u^{(1)}(\mathbf{x},\mathbf{y})\right) = -\nabla_\mathbf{y} \cdot \left(D(\mathbf{x},\mathbf{y})\nabla_\mathbf{x} u^{(0)}(\mathbf{x})\right) \text{ in } \Omega, \tag{1.44}$$

The above problem reads as a periodic cell problem for the first order coefficient $u^{(1)}(\mathbf{x},\mathbf{y})$ and it is once again a classical diffusion-type problem, equipped with a volume load on the right-hand side and periodic boundary conditions on $\partial\Omega$. It admits a unique solution up to an arbitrary **y**-constant function $c(\mathbf{x})$. Since the problem is linear and the vector function $\nabla_\mathbf{x} u^{(0)}$ is **y**-constant, we state the following solution *ansatz*

$$u^{(1)}(\mathbf{x},\mathbf{y}) = \mathbf{a}(\mathbf{x},\mathbf{y}) \cdot \nabla_\mathbf{x} u^{(0)} + c(\mathbf{x}). \tag{1.45}$$

Relationship (1.45) is indeed the solution of the problem (1.44) provided that the vector **a** solves the following cell problem

$$\nabla_\mathbf{y} \cdot \left(D(\mathbf{x},\mathbf{y})\nabla_\mathbf{y}\mathbf{a}\right) = -\nabla_\mathbf{y} D(\mathbf{x},\mathbf{y}) \text{ in } \Omega, \tag{1.46}$$

where **a** is **y**-periodic and a further condition is needed to achieve uniqueness, for example by fixing the integral average of **a** over the periodic cell Ω. As for the one-dimensional case, we need one last step to obtain a closed macroscale problem for the leading order coefficient $u^{(0)}$.

$\boxed{\epsilon^2}$

We are now ready to conclude the multiscale procedure by equating the same powers of ϵ^2 in (1.39), that yields

$$\nabla_\mathbf{x} \cdot \left(D\nabla_\mathbf{x} u^{(0)}\right) + \nabla_\mathbf{x} \cdot \left(D\nabla_\mathbf{y} u^{(1)}\right) + \\ \nabla_\mathbf{y} \cdot \left(D\nabla_\mathbf{x} u^{(1)}\right) + \nabla_\mathbf{y} \cdot \left(D\nabla_\mathbf{y} u^{(2)}\right) = f \tag{1.47}$$

We now average relationship (1.47) over the periodic cell, i.e. we apply the following cell average operator:

$$\langle (\bullet) \rangle_\Omega = \frac{1}{|\Omega|} \int_\Omega (\bullet) d\mathbf{y}, \qquad (1.48)$$

where $|\Omega|$ is the volume (or area, in two dimensions) of the periodic cell. Application of (1.48) to (1.47) yields

$$\nabla_\mathbf{x} \cdot \left(\langle D \rangle_\Omega \nabla_\mathbf{x} u^{(0)} \right) + \nabla_\mathbf{x} \cdot \langle D \nabla_\mathbf{y} u^{(1)} \rangle_\Omega +$$
$$\frac{1}{|\Omega|} \int_{\partial\Omega} D \nabla_\mathbf{x} u^{(1)} \cdot \mathbf{n} \, dS + \frac{1}{|\Omega|} \int_{\partial\Omega} D \nabla_\mathbf{y} u^{(2)} \cdot \mathbf{n} \, dS = \langle f \rangle_\Omega, \qquad (1.49)$$

where the surface integrals arise after applying the divergence theorem with respect to $\mathbf{y}$. Both the surface contributions actually read as integrals of the scalar product between a periodic function and the unit vector $\mathbf{n}$ normal to $\partial\Omega$ over their corresponding periodic cell. Accounting for periodicity, these terms identically reduce to zero, as every contribution on a single face of the periodic cell (edge in two-dimensions), is exactly canceled by the contribution on its corresponding parallel face considering the change in sign of the unit outward normal vector $\mathbf{n}$ (for example, in Fig. 1.4, the contribution over side A exactly cancels the one over side C, and the same holds for sides B and D).

Equation (1.49) can be further rearranged accounting for ansatz (1.45) obtaining:

$$\nabla_\mathbf{x} \cdot \left(\langle D \rangle_\Omega \nabla_\mathbf{x} u^{(0)} \right) + \nabla_\mathbf{x} \cdot \left(\langle D (\nabla_\mathbf{y} \mathbf{a})^\mathsf{T} \rangle_\Omega \nabla_\mathbf{x} u^{(0)} \right) = \langle f \rangle_\Omega. \qquad (1.50)$$

Finally, (1.50) can be rewritten as a macroscale diffusion problem for $u^{(0)}$ as follows:

$$\nabla_\mathbf{x} \cdot \left(\mathsf{D}(\mathbf{x}) \nabla_\mathbf{x} u^{(0)} \right) = \hat{f}(\mathbf{x}), \qquad (1.51)$$

equipped with appropriate macroscale boundary conditions, for example of the type (1.38). The homogenized diffusion tensor D and the macroscale volume source $\hat{f}$ are defined as

$$\mathsf{D}(\mathbf{x}) = \langle D \rangle_\Omega \mathsf{I} + \langle D (\nabla_\mathbf{y} \mathbf{a})^\mathsf{T} \rangle_\Omega, \qquad (1.52)$$

or, componentwise

$$D_{ij}(\mathbf{x}) = \langle D \rangle_\Omega \delta_{ij} + \left\langle D \frac{\partial a_j}{\partial y_i} \right\rangle_\Omega \qquad (1.53)$$

and

$$\hat{f}(\mathbf{x}) = \langle f \rangle_\Omega. \qquad (1.54)$$

where I is the identity tensor. The homogenized problem (1.51) is to be solved on the macroscale only, and microscale information is encoded in the components of the effective diffusivity tensor (1.52), which can be computed solving the diffusion-type cell problems given by (1.46) and exploiting (1.52). A few remarks and exercises now follow.

Remark 1.2 (Anisotropy) The homogenized problem reads as an anisotropic diffusion problem in the limit $\epsilon \to 0$. Hence, the microscale inhomogeneities characterizing the physical diffusion coefficient $D(\mathbf{x}, \mathbf{y})$ translate into anisotropy on the macroscale. In particular, the degree of anisotropy is in general related to the specific form of the coefficient $D(\mathbf{x}, \mathbf{y})$, that dictates the shape and relative dimension of the periodic cell where the vector **a**, and, in turn the components of the tensor D are to be computed.

Remark 1.3 (Computational Feasibility) Let us consider a diffusion coefficient of the type $D(\mathbf{y})$. Then, the cell problem (1.46) solely depends on the microscale variable **y** and can be solved once, independently from the macroscale **x**. In this case, replacing the original problem (1.37) with the homogenized problem (1.51) greatly reduces computational complexity. Given the solution to (1.46), it is straightforward to compute the effective diffusivity tensor (1.52) and finally solve the classical, homogeneous diffusion problem (1.51) on a coarse grid which captures macroscopic variations of the fields only. Whenever the coefficient D retains a macroscopic variation, then it is in principle necessary to solve one cell problem for every macroscale point **x**. However, since the macroscale domain is supposed to be represented by a coarse grid, computing a limited number of diffusion-type prescribed cell problems is, in most cases, still more advantageous than resolving the full microscale variations embedded in the original inhomogeneous diffusion problem.

Remark 1.4 (On the Role of Periodicity) We have carried out the asymptotic homogenization steps for the n-dimensional diffusion problem assuming periodicity of the microscale (cf. Assumption V) instead of local boundedness. As we remarked at the beginning of this section, this choice is primarily motivated by practical reasons, as the periodicity assumption enabled us to reconstruct microscopic information focusing on a limited portion of the microstructure, namely, the periodic cell. However, we would like to remark that this assumption is not necessary to derive the homogenized problem, and the analytic form of the microstructural problem as such, as everything could have been carried out assuming local boundedness only. In this case, the asymptotic homogenization technique serves as a powerful tool to derive reliable macroscale problems that can be used to model appropriate physical scenarios of interest, without computing the coefficients themselves. In this case, the latter are supposed to be obtained via other sources, for example experimental measurements. The reader could, as an exercise, derive the effective governing equations for the n-dimensional diffusion problem assuming local boundedness only, as done for the derivation of the equation of poroelasticity in [7].

Remark 1.5 (Macroscopic Uniformity) We implicitly assumed the so-called *macroscopic uniformity*, i.e. the periodic cell is independent of the macroscale. This assumption, which is often assumed implicitly in the asymptotic homogenization literature, allowed us to derive (1.49) assuming

$$\langle \nabla_{\mathbf{x}} \cdot (\bullet) \rangle_\Omega = \nabla_{\mathbf{x}} \cdot \langle (\bullet) \rangle_\Omega . \tag{1.55}$$

Whenever $\Omega = \Omega(\mathbf{x})$, relationship (1.55) does not hold, and proper application of the generalized Reynold's transport theorem is to be enforced to obtain additional macroscale volume sources that modify the homogenized diffusion problem (see, e.g. [13, 19, 20] and alternative approaches concerning multiscale definition of the unit normal vector for non macroscopically uniform domains, such as [6, 11, 12]). Furthermore, whenever the periodic cell retains a parametric dependence on the macroscale variable $\mathbf{x}$, the problem requires the solution of a periodic cell problem for each macroscale point $\mathbf{x}$, as we observed for the case of macroscopically varying diffusion coefficients D.

We conclude this section proposing the following exercises.

Exercise 1.1 Assume $n = 1$ and $D(x, y) = D(x, y + y_p)$. Solve the cell problem analytically in such a particular case and prove that

$$\bar{D}(x) = \left(\frac{1}{y_p} \int_0^{y_p} \frac{1}{D(x, s)} \, ds \right)^{-1}, \tag{1.56}$$

that is exactly the periodic counterpart of the relationship (1.36) derived in Sect. 1.2.

Exercise 1.2 Assume $n = 2$, $\mathbf{y} = (y_1, y_2)$ and $D(\mathbf{x}, \mathbf{y}) = D_0(\mathbf{x}) D_A(y_1) D_B(y_2)$, with $D_A(y_1) = D_A(y_1 + a)$ and $D_B(y_2) = D_B(y_2 + b)$. Solve the cell problem analytically and prove the following relationships for the components of the resulting homogenized diffusivity tensor D

$$D_{12} = D_{21} = 0, \tag{1.57}$$

$$D_{11} = \left(\frac{1}{a} \int_0^a \frac{1}{D_A(s)} \, ds \right)^{-1} \frac{1}{b} \int_0^b D_B(s) \, ds, \tag{1.58}$$

$$D_{22} = \left(\frac{1}{b} \int_0^b \frac{1}{D_B(s)} \, ds \right)^{-1} \frac{1}{a} \int_0^a D_A(s) \, ds. \tag{1.59}$$

The reader is invited to explore several possible variations of the diffusion coefficient $D(\mathbf{x}, \mathbf{y})$ and investigate how these affects the resulting components of the homogenized tensor D.

1.4 Porous Media Flow: Homogenization of the Stokes' Problem

The last introductory example we present concerns fluid flow in porous media. These materials are typically involved when dealing with several physical scenarios of practical interest, such as fluid flow through sand and rocks, and interstitial flow through biological tissues, for example bone, cell aggregates, organs and tumors. Here we analyze a simple, yet paradigmatic case, that is, the interaction between a solid rigid phase and an incompressible Newtonian fluid slowly flowing through the pores. We identify the whole physical domain with the open set $\Omega \subset \mathbb{R}^3$, $\bar{\Omega} = \bar{\Omega}_f \cup \bar{\Omega}_s$, where Ω_f and Ω_s are the fluid and solid regions, respectively. The fluid flow through the pores is then governed by the Stokes' problem, i.e.

$$\mu \nabla_\mathbf{x}^2 \mathbf{v} = \nabla p, \ \mathbf{x} \in \Omega_f \quad (1.60)$$

$$\nabla_\mathbf{x} \cdot \mathbf{v} = 0, \ \mathbf{x} \in \Omega_f \quad (1.61)$$

$$\mathbf{v} = \mathbf{0}, \text{ on } \Gamma, \quad (1.62)$$

where $\mathbf{v}$ is the fluid velocity, p the fluid pressure, and $\Gamma = \partial \Omega_f \cap \partial \Omega_s$ represents the interface between the two phases. Equations (1.60–1.62) represent the fluid stress balance, the incompressibility constraint and the no slip conditions for a low Reynolds number Newtonian incompressible fluid, respectively. We aim, once again, at obtaining a macroscale representation for such a problem, which is in this case particularly well suited, as the three dimensional porous structure could be, in general, extremely complex and the problem (1.60–1.62) practically impossible to solve also with numerical techniques. In particular, the sharp length scale separation in such a system relies on its geometry, rather than the analytic form of rapidly varying coefficients. In fact we can identify our microscale d with the pore *radius* (or an equivalent, average linear measure for non-cylindrical pores), and our macroscale with the average (linear) size of the whole domain, or, equivalently, with the average length of the pores. A sketch of the porous microstructure is provided in Fig. 1.5.

Fig. 1.5 The pore microstructure (shown on the left) against the macrostructure, where the geometrical variations are smoothed out (shown on the right)

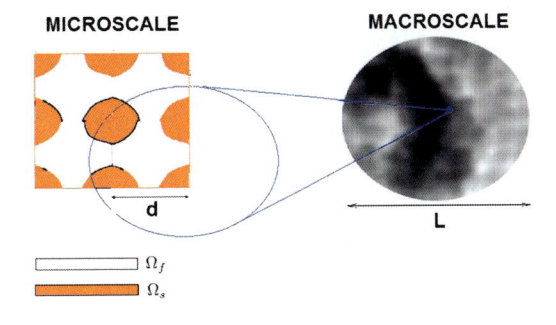

In this case, the multiscale nature of the problem is clearly dictated by the geometry itself, and it is necessary to perform an explicit non-dimensionalization process to fully account for the scale separation that characterizes the system.

1.4.1 Non-Dimensionalisation

We rescale our relevant fields as follows

$$\mathbf{x} = L\mathbf{x}', \quad \mathbf{v} = \frac{Cd^2}{\mu}\mathbf{v}', \quad p = CLp' \tag{1.63}$$

where C denotes the magnitude of a characteristic pressure gradient. Here, we scale the spatial coordinate by the characteristic size of the domain (pore length) L, whereas the characteristic velocity V is suggested by the parabolic profile of a viscous fluid flowing in a straight cylindrical channel of radius d, i.e.

$$V \propto \frac{Cd^2}{\mu}, \tag{1.64}$$

see classic fluid-dynamics textbooks, such as [9]. Since differential operators transform as

$$\nabla_{\mathbf{x}}^2 = \frac{1}{L^2}\nabla_{\mathbf{x}'} \tag{1.65}$$

and

$$\nabla_{\mathbf{x}} = \frac{1}{L}\nabla_{\mathbf{x}'}, \tag{1.66}$$

the non-dimensional Stokes' problem reads, in terms of the non-dimensional quantities (1.63) and neglecting the primes for the sake of simplicity of notation:

$$\epsilon^2 \nabla_{\mathbf{x}}^2 \mathbf{v} = \nabla p, \ \mathbf{x} \in \Omega_f \tag{1.67}$$

$$\nabla_{\mathbf{x}} \cdot \mathbf{v} = 0, \ \mathbf{x} \in \Omega_f \tag{1.68}$$

$$\mathbf{v} = \mathbf{0}, \text{ on } \Gamma, \tag{1.69}$$

where we recall that $\epsilon = d/L$.

1.4.2 The Homogenized Problem

We are dealing with a porous medium, and therefore assume that the average pore radius d is much smaller than the average size of the domain L, such that $\epsilon \ll 1$. We then apply the asymptotic homogenization Assumptions I to III, together with the local periodicity and regularity Assumptions IV and V (exploited in the same way as we have done for the n-dimensional diffusion problem) to the non-dimensional Stokes' problem. We further multiply both the right and the left hand sides of (1.67–1.69) by suitable powers of ϵ to obtain

$$\epsilon^3 \nabla_x^2 \mathbf{v}^\epsilon + \epsilon^2 \nabla_x \cdot \left(\nabla_y \mathbf{v}^\epsilon\right) + \epsilon^2 \nabla_y \cdot \left(\nabla_x \mathbf{v}^\epsilon\right) + \epsilon \nabla_y^2 \mathbf{v}^\epsilon = \nabla_y p^\epsilon + \epsilon \nabla_x p^\epsilon, \quad (1.70)$$

$$\nabla_y \mathbf{v}^\epsilon + \epsilon \nabla_x \mathbf{v}^\epsilon = 0 \quad (1.71)$$

in Ω_f and

$$\mathbf{v}^\epsilon = \mathbf{0} \quad (1.72)$$

on Γ.

We now equate the same powers of ϵ in ascending order from ϵ^0 in each of the Stokes' problem equations (1.70–1.72). Since we are in a periodic setting, we identify Ω_f and Ω_s with the corresponding fluid and solid phase within the periodic cell, which we call Ω.

$\boxed{\epsilon^0}$

Equating the same powers of ϵ^0 in the stress balance equation (1.70) yields

$$\nabla_y p^{(0)}(\mathbf{x}, \mathbf{y}) = 0 \Rightarrow p^{(0)} = p^{(0)}(\mathbf{x}), \quad (1.73)$$

that is, the leading order pressure depends only on the macroscale $\mathbf{x}$. The ϵ^0 conditions arising from the incompressibility constraint (1.71) and the no slip condition (1.72) read

$$\nabla_\mathbf{y} \cdot \mathbf{v}^{(0)} = 0, \quad \text{in } \Omega_f \quad (1.74)$$

and

$$\mathbf{v}^{(0)} = \mathbf{0}, \quad \text{on } \Gamma \quad (1.75)$$

respectively.

$\boxed{\epsilon^1}$

Equating the same powers of ϵ in (1.70–1.72) leads to the following conditions

$$\nabla_y^2 \mathbf{v}^{(0)} = \nabla_y p^{(1)} + \nabla_x p^{(0)} \quad \text{in } \Omega_f, \tag{1.76}$$

$$\nabla_y \cdot \mathbf{v}^{(1)} + \nabla_x \cdot \mathbf{v}^{(0)} = \mathbf{0} \quad \text{in } \Omega_f, \tag{1.77}$$

and

$$\mathbf{v}^{(1)} = 0, \quad \text{on } \Gamma. \tag{1.78}$$

We now exploit the conditions obtained by equating the same powers of ϵ to close the macroscale problem for the leading order fields $\mathbf{v}^{(0)}$ and $p^{(0)}$.

We collect conditions (1.76), (1.74) and (1.75) together to obtain the following auxiliary Stokes' problem for the fields $(\mathbf{v}^{(0)}, p^{(1)})$

$$\nabla_y^2 \mathbf{v}^{(0)} = \nabla_y p^{(1)} + \nabla_x p^{(0)} \quad \text{in } \Omega_f, \tag{1.79}$$

$$\nabla_y \cdot \mathbf{v}^{(0)} = 0, \quad \text{in } \Omega_f \tag{1.80}$$

$$\mathbf{v}^{(0)} = \mathbf{0}, \quad \text{on } \Gamma, \tag{1.81}$$

supplemented by $\mathbf{y}$-periodicity on the external boundary of the cell $\partial \Omega_f \setminus \Gamma$. We now exploit linearity of the system (1.79–1.81) and the fact that, according to (1.73), the leading order pressure $p^{(0)}$ depends on the macroscale only, to formulate the following ansatz for the solution

$$\mathbf{v}^{(0)} = -\mathbf{W} \nabla_x p^{(0)}, \tag{1.82}$$

$$p^{(1)} = -\mathbf{P} \cdot \nabla_x p^{(0)} + \bar{p}(\mathbf{x}). \tag{1.83}$$

The above expressions represent the unique (up to a $\mathbf{y}$-constant arbitrary function $\bar{p}(\mathbf{x})$) solution of the auxiliary Stokes' problem (1.79–1.81), provided that the auxiliary second rank tensor $\mathbf{W}$ and vector $\mathbf{P}$ solve the following Stokes'-type periodic cell problem

$$\nabla_y^2 \mathbf{W}^\mathsf{T} = \nabla_y \mathbf{P} - \mathbf{I} \quad \text{in } \Omega_f, \tag{1.84}$$

$$\nabla_y \cdot \mathbf{W} = 0 \quad \text{in } \Omega_f, \tag{1.85}$$

$$\mathbf{W} = 0, \quad \text{on } \Gamma, \tag{1.86}$$

The differential problem (1.84–1.86) is closed by $\mathbf{y}$-periodic conditions on $\partial \Omega_f \setminus \Gamma$ and a further condition on the auxiliary vector $\mathbf{P}$ to ensure the solution uniqueness, for example

$$\langle \mathbf{P} \rangle_{\Omega_f} = 0. \tag{1.87}$$

The problem (1.84–1.86) explicitly reads, by components

$$\frac{\partial W_{ij}}{\partial y_k \partial y_k} = \frac{\partial P_i}{\partial y_j} - \delta_{ij} \quad \text{in } \Omega_f, \tag{1.88}$$

$$\frac{\partial W_{ij}}{\partial y_j} = 0 \quad \text{in } \Omega_f, \tag{1.89}$$

$$W_{ij} = 0, \quad \text{on } \Gamma, \tag{1.90}$$

where $i, j, k = 1, 2, 3$ and sum over repeated indices is understood. Thus, the auxiliary Stokes-type problem (1.84–1.86) requires the solution of three standard Stokes' problem for every fixed $i = 1, 2, 3$. The three Stokes' periodic cell problems differ in the volume load, that is $\mathbf{e}_1, \mathbf{e}_2, \mathbf{e}_3$, for $i = 1, 2, 3$, respectively. Integral average of the solution ansatz (1.82) over the fluid domain leads to the macroscale governing equation relating the leading order velocity and pressure, namely

$$\langle \mathbf{v}^{(0)} \rangle_{\Omega_f} = - \langle \mathbf{W} \rangle_{\Omega_f} \nabla_x p^{(0)}, \tag{1.91}$$

i.e., the fluid flow is described by the Darcy's law on the macroscale domain. Hence, since the average leading order velocity can be computed via (1.91), we just need one more scalar equation for the leading pressure. We consider the integral average of relationship (1.77) and apply the divergence theorem with respect to the microscale variable $\mathbf{y}$ to obtain:

$$\frac{1}{|\Omega_f|} \int_{\partial \Omega_f / \Gamma} \mathbf{v}^{(1)} \cdot \mathbf{n}_{\Omega_f} \, dS + \frac{1}{|\Omega_f|} \int_{\Gamma} \mathbf{v}^{(1)} \cdot \mathbf{n}_\Gamma \, dS + \nabla_x \cdot \langle \mathbf{v}^0 \rangle_{\Omega_f} = 0, \tag{1.92}$$

where $\mathbf{n}_\Gamma$ and $\mathbf{n}_{\partial \Omega_f}$ are the unit outward vectors normal to Γ and $\partial \Omega_f \setminus \Gamma$, respectively. Since the contributions over the external boundary of Ω_f cancel out because of $\mathbf{y}$-periodicity and Eq. (1.78) holds on Γ, no surface contribution remains. The partial differential equation for the leading order pressure $p^{(0)}$ then reads as an effective divergence-free constraint for the average fluid velocity, that is:

$$\nabla_x \cdot \langle v^{(0)} \rangle_{\Omega_f} = -\nabla_x \cdot \left(\langle W \rangle_{\Omega_f} \nabla_x p^{(0)} \right) = 0. \tag{1.93}$$

Therefore, asymptotic homogenization of the Stokes' problem for porous media flow leads to the incompressible Darcy's law for the average fluid velocity. The effective, non-dimensional hydraulic conductivity is given by the tensor $\langle \mathbf{W} \rangle_{\Omega_f}$ which can be computed by solving the Stokes'-type periodic cell problem (1.84–1.86) on the periodic cell Ω. In this case, as long as macroscopic uniformity is assumed (i.e. $\Omega = \Omega(\mathbf{y})$), the effective hydraulic conductivity is homogeneous and solely depends on the geometry of the cell, which is in turn representative of the porous medium structure.

Remark 1.6 It is well-known that a Darcy's law of the type (1.91) can be experimentally verified also for heterogeneous, nonperiodic, porous microstructure. In fact, the Darcy's law for porous media flow can be derived by mixture theory (see, e.g., [21]), where no periodicity is assumed, and also in the context of asymptotic homogenization, assuming local boundedness only. However, local periodicity enables us to derive *computationally feasible* microscale problems that can be solved in practice on a small portion of the microstructure, as done for example in [17] in the context of tumor blood transport. This way, Darcy's law does not play merely the role of the effective set of governing equations for the fluid flow, but also encodes precise information concerning the geometry of the porous structure.

Remark 1.7 (Geometric Homogenization) Note that this example shows that the asymptotic homogenization technique can be carried out also for physical systems that are not characterized by fine scale variations of the coefficients. We have started from the Stokes' problem at constant viscosity and have finally obtained Darcy's law exploiting the sharp length scale separation that exists in the *geometry* itself, which is captured via an explicit non-dimensionalization analysis. In the most general case, physical systems can exhibit both fine scale variations of the coefficients and geometric heterogeneities. For example, when dealing with elastic composite materials (see, e.g. [14, 22] and recently developed computational analysis such as [18]), both oscillations of the elastic coefficients within a single elastic phase and the difference between different phases may be observed on the fine scale, and the two contributions lead in general to distinct contributions that appear in the corresponding cell problems that are to be computed to determine the effective elasticity tensor.

1.5 Concluding Remarks

We have presented a brief introduction to the asymptotic homogenization technique. The material is intended to serve as a first step to foster the curiosity of students and scientists approaching the topic for the first time. We have applied the technique to simple examples, such as the diffusion problem and the Stokes' problem for porous media flow. These are only partially representative of the whole realm of multiscale, multiphysics problems and have been chosen to drive the reader's attention towards the fundamental significance of spatial scale decoupling and non-dimensionalization, and the importance of appropriate regularity assumption (local boundedness, local periodicity) in deriving appropriate homogenized PDEs. We have deliberately ignored advanced, cutting edge applications, as this book chapter solely serves as a simple, basic introduction to the topic. However, we believe that this introductory work may help the interested readers to understand fundamental issues concerning the technique and to raise their awareness when facing complex multiscale problems involving the interplay among several physical phenomena.

Acknowledgements R.P. conceived and wrote the present book chapter during his previous appointment at TU Darmstadt, where he was supported by the DFG priority program SPP 1420, project GE 1894/3 and RA 1380/7, Multiscale structure-functional modeling of musculoskeletal mineralized tissues, PIs Alf Gerisch and Kay Raum.

References

1. Allaire G (1992) Homogenization and two-scale convergence. SIAM J Math Anal 23(6):1482–1518
2. Auriault JL, Boutin C, Geindreau C (2010) Homogenization of coupled phenomena in heterogenous media, vol 149. Wiley, New York
3. Bakhvalov N, Panasenko G (1989) Homogenisation averaging processes in periodic media. Springer, New York
4. Bowen R (1980) Incompressible porous media models by the use of the theory of mixtures. Int J Eng Sci 18:1129–1148
5. Bowen R (1982) Compressible porous media models by the use of the theory of mixtures. Int J Eng Sci 20:697–735
6. Bruna M, Chapman SJ (2015) Diffusion in spatial varying porous media. SIAM J Appl Math 75(4):1648–1674
7. Burridge R, Keller J (1981) Poroelasticity equations derived from microstructure. J Acoust Soc Am 70:1140–1146
8. Cherkaev A, Kohn R (1997) Topics in the mathematical modelling of composite materials. Springer, New York
9. Chorin AJ, Marsden JE (1993) A mathematical introduction to fluid dynamics. Springer, New York
10. Cioranescu D, Donato P (1999) An introduction to homogenization. Oxford University Press, Oxford
11. Dalwadi MP (2018) Asymptotic homogenization with a macroscale variation in the microscale. In: Gerisch A, Penta R, Lang J (eds) Multiscale models in mechano and tumor biology: modeling, homogenization, and applications. Lecture notes in computational science and engineering, chap 2. Springer, Heidelberg
12. Dalwadi MP, Griffiths IM, Bruna M (2015) Understanding how porosity gradients can make a better filter using homogenization theory. Proc R Soc A 471(2182):20150,464
13. Holmes M (1995) Introduction to perturbation method. Springer, New York
14. Mei CC, Vernescu B (2010) Homogenization methods for multiscale mechanics. World Scientific, Singapore
15. Murat F (1978) H-Convergence, Séminaire d'Analyse Fonctionnelle et Numérique (1977/1978). Université d'Alger, Multigraphed
16. Papanicolau G, Bensoussan A, Lions JL (1978) Asymptotic analysis for periodic structures. Elsevier, Amsterdam
17. Penta R, Ambrosi D (2015) The role of microvascular tortuosity in tumor transport phenomena. J Theor Biol 364:80–97
18. Penta R, Gerisch A (2016) Investigation of the potential of asymptotic homogenization for elastic composites via a three-dimensional computational study. Comput Vis Sci 17(4):185–201
19. Penta R, Ambrosi D, Shipley RJ (2014) Effective governing equations for poroelastic growing media. Q J Mech Appl Math 67(1):69–91
20. Penta R, Ambrosi D, Quarteroni A (2015) Multiscale homogenization for fluid and drug transport in vascularized malignant tissues. Math Models Methods Appl Sci 25(1):79–108

21. Rajagopal K (2007) On a hierarchy of approximate models for flows of incompressible fluids through porous solids. Math Models Methods Appl Sci 17(02):215–252
22. Sanchez-Palencia E (1980) Non-homogeneous media and vibration theory. Springer, New York
23. Shipley RJ, Chapman J (2010) Multiscale modelling of fluid and drug transport in vascular tumors. Bull Math Biol 72:1464–1491

Chapter 2
Asymptotic Homogenization with a Macroscale Variation in the Microscale

Mohit P. Dalwadi

2.1 Introduction

Asymptotic homogenization can often be used to reduce the complexity of a problem that has a periodic geometry. Starting from the governing equations for the full problem, the general idea behind asymptotic homogenization is to obtain governing equations for the variables averaged over one periodic cell, and this length scale is referred to as the *microscale*. Determining the homogenized equations usually requires solving a given problem over one periodic cell, known as the *cell problem*. Solving the cell problem once and then solving the resulting homogenized equations is generally less computationally expensive than solving the full problem. This is because the homogenization procedure has removed the periodic variation from the problem while retaining the slow change over many periodic cells, and this longer length scale is referred to as the *macroscale*. Asymptotic homogenization methods have been widely studied in the literature (see, for example, [2, 10, 11, 18, 19]).

We will consider asymptotic homogenization via the method of multiple scales [2] rather than, for example, volume averaging methods [19]. Generally, the assumptions required to apply the technique of asymptotic homogenization via the method of multiple scales are: (i) there is a periodic microstructure and (ii) the ratio between the length of the periodic cell and the length of the macroscale variation is small. An introduction to this method is given in Chap. 1. In this chapter, we see how to relax the assumption that the microstructure is strictly periodic, and we consider a microscale structure that varies over the macroscale.

M.P. Dalwadi (✉)
Synthetic Biology Research Centre, University of Nottingham, NG7 2RD Nottingham, UK
e-mail: mohit.dalwadi@nottingham.ac.uk

2.1.1 Literature Review

The method we consider in this chapter to deal with a macroscopic variation in the microstructure has been applied to a wide variety of problems (see, for example, [3, 7–9, 12–17]), and has formal analysis roots in, for example, [1, 5]. The general idea behind this method is to prescribe a level-set function to define the microstructure in both the microscale and macroscale variables. To highlight the general method and a computationally efficient reduction, we now discuss [15, 17], and [3] in more detail.

In [17], the authors consider the problem of deriving homogenized equations to determine the electric potential within a beating heart, where the time-dependent microstructure is close to spatially periodic in general curvilinear coordinates. Thus, the microscale may vary spatially over the macroscale. The authors transform the microscale to a strictly periodic domain, and then perform an asymptotic homogenization via the method of multiple scales on the transformed problem. The transformation of the microscale means that the derived cell problem involves coefficients from the Jacobian matrix of the transformation. Therefore, unlike a problem with strict periodicity in the microscale (such as those considered in Chap. 1), a different cell problem must now be solved at every point in the macroscale rather than just once for the entire problem. Although this procedure is less computationally expensive than solving the full problem, there is still a significant computational expense associated with solving many different cell problems.

In [15], the dynamics of colloids suspended within a fluid moving past circular obstacles in a square array are homogenized and cell problems are derived. The colloidal particles are allowed to diffuse, advect, interact with one another due to a general interaction potential, and are allowed to attach and detach to and from the obstacles. Thus, the microstructure is not strictly periodic and the level set method is used to homogenize the problem. Although the obstacle variation is constrained to a one-parameter family, the strong coupling between the flow and colloid particle problems means that a different cell problem must be solved at every point in the macroscale for each time step, and there remains an appreciable computational expense.

One way to reduce the cost of solving many different cell problems is to constrain the problem so that there are a limited number of possible cell problems to solve. This is the route taken in [3], where the authors consider the problem of deriving homogenized equations to the diffusion equation in a domain obstructed by nonoverlapping impermeable spheres whose centres are located on a periodic lattice. The radius of adjacent spheres is allowed to vary by an amount that is small compared to the separation of sphere centres. Defining the separation of sphere centres as the microscale length and the slow variation of sphere radius as the macroscale length, this set-up allows for a macroscale variation in microstructure. Importantly, this variation is constrained to a one-parameter family: the sphere radius. Thus, although one must still solve several cell problems to yield the homogenized problem, the range of possible cell problems only involves varying one parameter. Therefore, the general homogenized problem can be fully characterized by solving, say, 50 cell problems, each using a different sphere radius, and interpolating the relevant data for a sphere whose radius falls between two of the calculated data points. As the sphere radius determines the solid fraction of a cell, the entire homogenized problem can be written in terms of the cell-averaged porosity. An interesting conclusion from [3] is that a porosity variation induces a macroscale advection of concentration averaged over the entire cell in the direction of decreasing porosity.

A notable difference between the main problem presented in [17] to those considered in [3, 15] is that the microstructure is strictly periodic in [17] after transformation, whereas the microstructure is only close to periodic in the latter two. Thus, the main cell problems in [17] vary due to the differing coefficients in the governing equations, whereas the main cell problems in [3, 15] vary due to their differing geometries. It is shown in [17] that one can move between these two formulations for a general transformation, and it is further shown in [3] that a conformal transformation has a simplified cell problem due to the form of the Jacobian matrix. Moreover, as noted in [3], a conformal transformation also has the property that spherical obstacles remain spherical, and thus it is relatively simple to switch between the cases of sphere centres being located on a strictly periodic lattice with varying sphere radius and sphere centres being located on a locally periodic lattice with constant sphere radius. In either case, particular care must be taken when evaluating the unit normal to the obstacle surface, which may appear in Neumann or Robin boundary conditions. Between two near-to-periodic cells, there may be a small variation in the unit normal to the obstacle surface, due to a change in the position of the surface. As emphasized in [3], this must be taken into account during the homogenization procedure.

An important modelling question is whether a regular lattice can be a good approximation of an unstructured medium that may be encountered in physical problems. This has been investigated in [6] and [3]. In [6], the steady problem of nutrient uptake past randomly placed point sinks is considered in one spatial dimension. As the governing equations can be solved if the locations of the sinks are known, significant analytic progress is made into investigating the macroscale effect of different random distributions. The authors show that, although leading-order approximations remain the same between periodic and random microscale structure, the error terms vary in magnitude and large spatial gradients can occur from error terms in certain parameter regimes. In [3], the authors also investigate the error introduced when one treats an unstructured microstructure as near-periodic. The authors compare results for an ordered and disordered microstructure in the limit of a low fraction of obstacles. They solve the full problem in both cases, and compare these results to the solution of the homogenized problem, concluding that there is little difference between the solutions.

The method used in [3] to homogenize the diffusion equation past impenetrable spheres has been extended to consider filtration problems in a similar domain. In [7], the authors homogenize the flow past a periodic array of impermeable spheres with a near-periodic microstructure and the coupled problem of solute transport owing to advection, diffusion, and adsorption onto the surfaces of the spheres. As in [3], the near periodicity of the spheres in [7] is due to a slow spatial variation in sphere radius. The motivation of [7] is to understand why filters with gradients in porosity tend to be more effective than uniform filters, where the spherical obstacles model the filter. The authors find that filtration is significantly more uniform in filters whose porosity decreases with depth compared to uniform filters, but the average filtration tends to be similar. As a large particulate removal in one place may result in reduced pore space, it is conjectured in [7] that filters with a decreasing porosity have a longer lifespan before blocking. This conjecture is confirmed in a subsequent paper [8], where a similar problem to [7] is considered, but now the blocking effect is explicitly accounted for by allowing the spheres to grow in time according to their adsorption of particulates. Thus, the microstructure now varies both temporally and spatially. Although the system presented in [8] involves a moving boundary as in [15], the problem is simplified by exploiting the slow growth of obstacles due to slow particle accumulation compared to flow velocity, thus the range of possible cell problems only involves varying one parameter as in [3]. Using asymptotic results, the authors are able to solve the inverse problem of determining the initial porosity distribution of filters that block everywhere at once, and they show that these filters remove more particulates than other filters with the same initial average porosity. The general framework presented in [8] could also consider shrinking obstacles in, for example, chemotherapy delivery, by changing the sign of the obstacle growth term.

2.1.2 Chapter Outline

In this chapter, we consider the problem of homogenizing the concentration field of a dissolved drug that diffuses within a two-dimensional domain containing circular inclusions. The boundaries of these circles can absorb the drug, and the circle centres are arranged on a periodic square lattice. We define distances of the same order as the distance between circle centres as the *microscale*, and we allow the sphere radii to vary spatially over a significantly larger distance that we term the *macroscale*. This set-up could model drug delivery to tissue, but we do not dwell on the physical implications of this as our main focus is on the method used to homogenize a microscale with a macroscale variation.

2.2 Model Set-Up

We consider a system where the concentration field of a dissolved drug evolves due to diffusion within a domain obstructed by tissue, modelled as a periodic square array of circular obstacles. The boundaries of this tissue act as sinks for

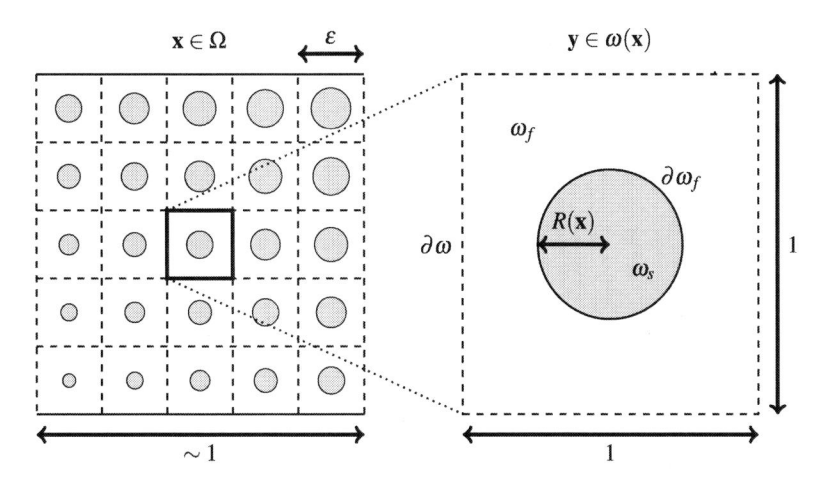

Fig. 2.1 A schematic of the model. Left: An example of a near-to-periodic set-up where the microscale varies over the macroscale. Right: A magnified view of a given cell $\omega(\mathbf{x})$, with microscale coordinate $\mathbf{y} \in \left[-\frac{1}{2}, \frac{1}{2}\right]^2$.

the concentration field, modelling drug delivery. We start with the dimensionless problem, to allow us to focus on the homogenization procedure.

The concentration field is given by $c(\mathbf{x}, t)$ (where $\mathbf{x}$ is the spatial vector coordinate and t is time), and is defined within $\Omega_f \subset \mathbb{R}^2$, outside the array of circular tissue.[1] Although we do not explicitly account for fluid flow in this chapter (this extension is considered in [7, 8, 15]), it is helpful to refer to Ω_f as the fluid phase. We define the tissue as $\Omega_s \subset \mathbb{R}^2$, and we refer to Ω_s as the solid phase. The entire domain is $\Omega = \Omega_f \cup \Omega_s \subset \mathbb{R}^2$, and we note that the fluid and solid phase are distinct, so that $\Omega_f \cap \Omega_s = \emptyset$. The circular boundaries between fluid and solid phase are defined as $\partial \Omega_f$. The tissue is modelled by a collection of fixed non-overlapping circles, whose centres are located on a square lattice at a distance ϵ apart, where ϵ is a small dimensionless parameter and the typical dimensionless macroscale length is 1. We allow the circle radii to vary in space, and a circle with centre at $\mathbf{x}$ has radius $\epsilon R(\mathbf{x})$, where $R = O(1)$. A schematic of this set-up is given in Fig. 2.1.

The concentration field is governed by the standard diffusion equation with a partially absorbing Robin boundary condition:

$$\frac{\partial c}{\partial t} = \nabla^2 c, \qquad \mathbf{x} \in \Omega_f, \tag{2.1a}$$

$$-\epsilon \gamma c = \mathbf{n} \cdot \nabla c, \qquad \mathbf{x} \in \partial \Omega_f, \tag{2.1b}$$

---

[1] We note that this method can easily be extended to three dimensions, as seen in [3, 7, 8].

and we are interested in the cases where $c = O(1)$ and $t = O(1)$. Physically, the boundary condition (2.1b) states that the solute uptake on the circular boundaries (equivalently, the flux of solute into the circular obstacle) is proportional to the concentration of the drug on the tissue boundary. Here, $\epsilon\gamma = O(\epsilon)$ is an experimentally determined uptake coefficient that will depend on the combination of drug and tissue. As the uptake coefficient is of $O(\epsilon)$, there is a small flux into each obstacle and this leads to a distinguished asymptotic limit in the final homogenized equation, where all mechanisms contribute at leading order. The boundary condition (2.1b) is a simple model of drug uptake on the tissue boundary, and the left-hand side can easily be generalised to a different uptake model, as long as the uptake coefficient is of $O(\epsilon)$. For example, one could alternatively use a Michaelis–Menten type condition $-\epsilon\gamma c/(K+c)$ on the left-hand side of (2.1b), where K is constant, to model saturation of uptake.

We now perform an asymptotic homogenization using the method of multiple scales on the problem defined by (2.1), exploiting the asymptotic limit $\epsilon \to 0$. This limit corresponds to there being a small ratio of distance between adjacent circle centres and the length scale of the macroscale variation in circle radius.

2.3 Homogenization

To perform the asymptotic homogenization, we introduce the microscale variable $\mathbf{y} = \mathbf{x}/\epsilon$ and treat $\mathbf{x}$ and $\mathbf{y}$ as independent. The extra degree of freedom introduced by this additional independent variable is later removed by imposing that the solution is exactly periodic in $\mathbf{y}$. Hence, any small variation between unit cells is captured through the macroscale variable $\mathbf{x}$. The microscale variable $\mathbf{y}$ is defined within a given unit cell $\omega(\mathbf{x})$, whereas the macroscale variable $\mathbf{x}$ is defined across the entire domain (Fig. 2.1). The tissue is denoted $\omega_s(\mathbf{x})$ and the fluid portion is $\omega_f(\mathbf{x}) = \omega(\mathbf{x}) \setminus \omega_s(\mathbf{x})$. The circular tissue–fluid boundary within the unit cell is denoted by $\partial\omega_f(\mathbf{x})$, while the square outer boundary of the unit cell is $\partial\omega(\mathbf{x})$.

We now consider each dependent variable as a function of both $\mathbf{x}$ and $\mathbf{y}$. Using the new variable $\mathbf{y}$, we transform spatial derivatives in the following manner

$$\nabla \mapsto \nabla_\mathbf{x} + \frac{1}{\epsilon}\nabla_\mathbf{y}, \qquad (2.2)$$

where $\nabla_\mathbf{x}$ and $\nabla_\mathbf{y}$ refer to the nabla operator in the $\mathbf{x}$- and $\mathbf{y}$-coordinate systems respectively.

As our eventual goal is to derive homogenized equations that are valid in the macroscale domain, it will be useful to introduce some quantities averaged over the microscale. Our eventual homogenized equations will be in terms of these averaged quantities. For this purpose, we define the cell-averaged porosity $\phi(\mathbf{x})$ to be

$$\phi(\mathbf{x}) = \frac{|\omega_f(\mathbf{x})|}{|\omega(\mathbf{x})|} = |\omega_f(\mathbf{x})|, \qquad (2.3)$$

where we use the fact that $|\omega(\mathbf{x})| = 1$ in our microscale geometry to obtain the second equality. In a different microscale, for example, a hexagonal lattice rather than a square one, the cell area may be a different constant. As $\phi = 1 - \pi R^2$, we note that the cell-averaged porosity is bounded above by $1 - \pi/4$, which occurs when a circle radius is 0.5. This results in adjacent circles touching, and a subsequent change in the topology of the domain, so we restrict our domain to $0 < R < 0.5$.

There are two main ways to describe an average concentration in our problem, depending on whether the averaging takes place over the fluid phase of the cell or over the entire cell. These different averaging methods are known as the intrinsic-averaged and the volumetric-averaged concentrations, respectively. Formally, the intrinsic-averaged concentration is defined as

$$\tilde{C}(\mathbf{x}, t) = \frac{1}{|\omega_f(\mathbf{x})|} \int_{\omega(\mathbf{x})} c(\mathbf{x}, \mathbf{y}, t)\, d\mathbf{y} = \frac{1}{\phi(\mathbf{x})} \int_{\omega_f(\mathbf{x})} c(\mathbf{x}, \mathbf{y}, t)\, d\mathbf{y}, \quad (2.4a)$$

and the volumetric-averaged concentration is defined as

$$C(\mathbf{x}, t) = \frac{1}{|\omega(\mathbf{x})|} \int_{\omega(\mathbf{x})} c(\mathbf{x}, \mathbf{y}, t)\, d\mathbf{y} = \int_{\omega_f(\mathbf{x})} c(\mathbf{x}, \mathbf{y}, t)\, d\mathbf{y}, \quad (2.4b)$$

imposing that $c \equiv 0$ in $\omega_s(\mathbf{x})$.

2.3.1 Transforming the Normal

A key consideration in homogenizing a problem whose microscale structure varies in the macroscale is the form of the unit normal. Thus, this is not an issue when a problem has Dirichlet boundary conditions. However, for Neumann or, as in our problem, Robin boundary conditions, we must take care with the unit normal to the tissue boundary. This is carried out by considering a level set function, as used in [9, 12, 13, 16] for a general obstacle shape, and in [3, 7, 8] for the specific case of a circular obstacle.

To derive the unit normal, we introduce the scalar function

$$\chi(\mathbf{x}, \mathbf{y}) = R(\mathbf{x}) - \|\mathbf{y}\|, \quad (2.5)$$

where $\chi(\mathbf{x}, \mathbf{y}) = 0$ defines the tissue–fluid interface in a cell. Then, the normal vector $\mathbf{n}$ in (2.1b) becomes

$$\mathbf{n} = \frac{\nabla \chi}{\|\nabla \chi\|} = \frac{\mathbf{n}_\mathbf{y} + \epsilon \nabla_\mathbf{x} R}{\|\mathbf{n}_\mathbf{y} + \epsilon \nabla_\mathbf{x} R\|}, \quad (2.6)$$

where the second equality arises from the gradient transform (2.2), $\mathbf{n}_\mathbf{y} = -\mathbf{y}/\|\mathbf{y}\|$ is the outward unit normal on the obstacle boundary $\partial \omega_f(\mathbf{x})$, and $\epsilon \nabla_\mathbf{x} R$ accounts for the macroscale effect of varying obstacle size. This latter term is unexpected, but we will see later that it is vital in tracking how adjacent cells vary.

Before we carry on with the homogenization procedure, we note that $\mathbf{x}$ varies by an $O(\epsilon)$ amount across one cell. We show in Appendix 1 that switching between a small variation in $\mathbf{x}$ across one cell and taking $\mathbf{x}$ to be constant within one cell does not affect our analysis. It should be noted, however, that this small variation is an issue if there is an integral constraint in the problem, even for a strictly periodic microscale [4].

2.3.2 Homogenization Procedure

Using the transformations (2.2) and (2.6), the solute-transport problem (2.1) in one cell is

$$\epsilon^2 \frac{\partial c}{\partial t} = \left(\nabla_\mathbf{y} + \epsilon \nabla_\mathbf{x}\right) \cdot \left(\left(\nabla_\mathbf{y} + \epsilon \nabla_\mathbf{x}\right) c\right), \qquad \mathbf{y} \in \omega_f(\mathbf{x}), \qquad (2.7a)$$

$$-\epsilon^2 \gamma c = \left(\mathbf{n_y} + \epsilon \nabla_\mathbf{x} R\right) \cdot \left(\left(\nabla_\mathbf{y} + \epsilon \nabla_\mathbf{x}\right) c\right) + O(\epsilon^3), \qquad \mathbf{y} \in \partial \omega_f(\mathbf{x}), \qquad (2.7b)$$

$$c \text{ periodic}, \qquad \mathbf{y} \in \partial \omega(\mathbf{x}). \qquad (2.7c)$$

The homogenization procedure entails investigating the system (2.7) by looking for an asymptotic solution in the limit as $\epsilon \to 0$. That is, we look for a solution to c in terms of an asymptotic expansion

$$c(\mathbf{x}, \mathbf{y}, t) = c_0(\mathbf{x}, \mathbf{y}, t) + \epsilon c_1(\mathbf{x}, \mathbf{y}, t) + \epsilon^2 c_2(\mathbf{x}, \mathbf{y}, t) + O(\epsilon^3), \qquad (2.8)$$

and use the method of multiple scales to derive solvability conditions for c_0. These will be in terms of the macroscale variable $\mathbf{x}$ and will yield the homogenized equations for which we are looking. Although our main interest is in c_0, we will see that we need to proceed to terms of $O(\epsilon^2)$ to derive the solvability conditions for c_0.

2.3.2.1 The $O(1)$ Problem

Substituting the asymptotic expansion (2.8) into the system (2.7) and equating powers of ϵ, the $O(1)$ terms yield the problem

$$0 = \nabla_\mathbf{y}^2 c_0, \qquad \mathbf{y} \in \omega_f(\mathbf{x}), \qquad (2.9a)$$

$$0 = \mathbf{n_y} \cdot \nabla_\mathbf{y} c_0, \qquad \mathbf{y} \in \partial \omega_f(\mathbf{x}), \qquad (2.9b)$$

$$c_0 \text{ periodic}, \qquad \mathbf{y} \in \partial \omega(\mathbf{x}). \qquad (2.9c)$$

The system (2.9) can be solved by a c_0 that is independent of $\mathbf{y}$, and the linearity of (2.9) allows us to deduce that this solution is unique. We can therefore deduce that c_0 is independent of the microscale variable, *i.e.* $c_0 = c_0(\mathbf{x}, t)$, and this will be useful in simplifying the systems that arise from the $O(\epsilon)$ and $O(\epsilon^2)$ problems.

2.3.2.2 The $O(\epsilon)$ Problem

Substituting the asymptotic expansion (2.8) into the system (2.7) and equating powers of ϵ, the $O(\epsilon)$ terms yield the problem

$$0 = \nabla_{\mathbf{y}}^2 c_1, \qquad \mathbf{y} \in \omega_f(\mathbf{x}), \tag{2.10a}$$

$$-\mathbf{n}_{\mathbf{y}} \cdot \nabla_{\mathbf{x}} c_0 = \mathbf{n}_{\mathbf{y}} \cdot \nabla_{\mathbf{y}} c_1, \qquad \mathbf{y} \in \partial\omega_f(\mathbf{x}), \tag{2.10b}$$

$$c_1 \text{ periodic}, \qquad \mathbf{y} \in \partial\omega(\mathbf{x}). \tag{2.10c}$$

Although we cannot solve the system (2.10) analytically, we can derive a solvability condition by integrating (2.10a) over $\omega_f(\mathbf{x})$ and using the boundary conditions (2.10b)–(2.10c). However, the solvability condition we obtain from this is just

$$-\nabla_{\mathbf{x}} c_0 \cdot \oint_{\partial\omega_f(\mathbf{x})} \mathbf{n}_{\mathbf{y}} \, ds = 0, \tag{2.11}$$

where ds denotes the differential element of the obstacle boundary $\partial\omega_f(\mathbf{x})$, and we are able to take c_0 outside the integral because it is independent of $\mathbf{y}$. As (2.11) is trivially satisfied for any closed obstacle, we are not yet able to form a macroscale equation for c_0.

It will be useful to determine c_1 from (2.10) for use in the solvability condition we derive in the $O(\epsilon^2)$ problem. We could do this numerically for any given function c_0, but this relies on us knowing c_0, which is the function for which we are trying to solve. Alternatively, we can note that we just require knowledge of how c_1 behaves as a function of c_0. To do this, we note that the system (2.10) is linear in c_1, and we can therefore write c_1 as the dot product of some vector function $\boldsymbol{\Gamma}$ to be determined, and $\nabla_{\mathbf{x}} c_0$, as follows

$$c_1(\mathbf{x}, \mathbf{y}, t) = -\boldsymbol{\Gamma}(\mathbf{x}, \mathbf{y}) \cdot \nabla_{\mathbf{x}} c_0(\mathbf{x}, t). \tag{2.12}$$

This allows us to reduce the problem of determining c_1 in terms of c_0 to solving the following cell problem

$$0 = \nabla_{\mathbf{y}}^2 \Gamma_i, \qquad \mathbf{y} \in \omega_f(\mathbf{x}), \tag{2.13a}$$

$$n_{y,i} = \mathbf{n}_{\mathbf{y}} \cdot \nabla_{\mathbf{y}} \Gamma_i, \qquad \mathbf{y} \in \partial\omega_f(\mathbf{x}), \tag{2.13b}$$

$$\Gamma_i \text{ periodic}, \qquad \mathbf{y} \in \partial\omega(\mathbf{x}), \tag{2.13c}$$

where Γ_i is the ith component of $\boldsymbol{\Gamma}$, and $n_{y,i}$ is the ith component of the unit vector $\mathbf{n_y}$. For the circular boundaries we are considering, we have $n_{y,i} = -y_i/R(\mathbf{x})$, where y_i is the ith component of the microscale variable $\mathbf{y}$. In practice, $\boldsymbol{\Gamma}$ can be determined by using a finite element software package.

2.3.2.3 The $O(\epsilon^2)$ Problem

In our homogenization calculations, we have not yet used the fact that the microstructure is *near* periodic rather than *strictly* periodic. This is because the $O(\epsilon)$ correction to the normal in (2.6) has not yet appeared in our calculations. In this section, the macroscale variation in the microstructure finally imposes an effect.

Substituting the asymptotic expansion (2.8) into the system (2.7) and equating powers of ϵ, the $O(\epsilon^2)$ terms yield the problem

$$\frac{\partial c_0}{\partial t} = \nabla_\mathbf{y} \cdot \left(\nabla_\mathbf{y} c_2 + \nabla_\mathbf{x} c_1 \right) + \nabla_\mathbf{x} \cdot \left(\nabla_\mathbf{y} c_1 + \nabla_\mathbf{x} c_0 \right), \qquad \mathbf{y} \in \omega_f(\mathbf{x}), \qquad (2.14a)$$

$$-\gamma c_0 = \mathbf{n_y} \cdot \left(\nabla_\mathbf{y} c_2 + \nabla_\mathbf{x} c_1 \right) + \nabla_\mathbf{x} R \cdot \left(\nabla_\mathbf{y} c_1 + \nabla_\mathbf{x} c_0 \right), \qquad \mathbf{y} \in \partial \omega_f(\mathbf{x}), \qquad (2.14b)$$

$$c_2 \text{ periodic}, \qquad \mathbf{y} \in \partial \omega(\mathbf{x}). \qquad (2.14c)$$

As with (2.10), we cannot solve the system (2.14) analytically. Rather, we are interested in obtaining a solvability condition from (2.14), and this will yield our homogenized equation for c_0.

We can derive a solvability condition by integrating (2.14a) over ω_f to obtain

$$\int_{\omega_f(\mathbf{x})} \frac{\partial c_0}{\partial t} \, \mathrm{d}\mathbf{y} = \oint_{\partial \omega_f(\mathbf{x})} \mathbf{n_y} \cdot \left(\nabla_\mathbf{y} c_2 + \nabla_\mathbf{x} c_1 \right) \, \mathrm{d}s + \int_{\omega_f(\mathbf{x})} \nabla_\mathbf{x} \cdot \left(\nabla_\mathbf{y} c_1 + \nabla_\mathbf{x} c_0 \right) \, \mathrm{d}\mathbf{y}, \tag{2.15}$$

where the first term on the right-hand side of (2.15) has been obtained by using the divergence theorem and applying the boundary condition (2.14c).

The integrand on the left-hand side of (2.15) is independent of $\mathbf{y}$, and so can be integrated immediately. The first term on the right-hand side of (2.15) can be turned into a function of c_0 and c_1 by using the boundary condition (2.14b). Thus, we can re-write (2.15) as

$$|\omega_f(\mathbf{x})| \frac{\partial c_0}{\partial t} = \int_{\omega_f(\mathbf{x})} \nabla_\mathbf{x} \cdot \left(\nabla_\mathbf{y} c_1 + \nabla_\mathbf{x} c_0 \right) \, \mathrm{d}\mathbf{y} - \oint_{\partial \omega_f(\mathbf{x})} \nabla_\mathbf{x} R \cdot \left(\nabla_\mathbf{y} c_1 + \nabla_\mathbf{x} c_0 \right) \, \mathrm{d}s$$

$$- \oint_{\partial \omega_f(\mathbf{x})} \gamma c_0 \, \mathrm{d}s, \tag{2.16}$$

where we have rearranged the terms on the right-hand side.

The last term on the right-hand side of (2.16) is independent of $\mathbf{y}$, and so can be integrated immediately. The first two terms on the right-hand side of (2.16) can be simplified using the transport theorem (discussed in Appendix 2), and we can therefore write (2.16) as

$$|\omega_f(\mathbf{x})|\frac{\partial c_0}{\partial t} = \nabla_\mathbf{x} \cdot \int_{\omega_f(\mathbf{x})} (\nabla_\mathbf{y} c_1 + \nabla_\mathbf{x} c_0)\, \mathrm{d}\mathbf{y} - \gamma|\partial \omega_f(\mathbf{x})|c_0. \quad (2.17)$$

Our solvability condition is almost solely in terms of c_0, as we require. The final step is to use the result (2.12) to write $\nabla_\mathbf{y} c_1$ in terms of c_0, deducing that

$$\nabla_\mathbf{y} c_1 = \mathbf{J}_\Gamma^T \nabla_\mathbf{x} c_0, \quad (2.18\mathrm{a})$$

where $\mathbf{J}_\Gamma^T$ is the transpose of the Jacobian matrix of Γ, which is given by

$$\mathbf{J}_\Gamma = \begin{pmatrix} \frac{\partial \Gamma_1}{\partial y_1} & \frac{\partial \Gamma_1}{\partial y_2} \\ \frac{\partial \Gamma_2}{\partial y_1} & \frac{\partial \Gamma_2}{\partial y_2} \end{pmatrix}, \quad (2.18\mathrm{b})$$

where Γ can be obtained by solving the cell problem (2.13).

We can use the fact that $c_0 = c_0(\mathbf{x}, t)$ and (2.18) to write (2.17) as

$$|\omega_f(\mathbf{x})|\frac{\partial c_0}{\partial t} = \nabla_\mathbf{x} \cdot \left(\left(\int_{\omega_f(\mathbf{x})} (\mathbf{I} - \mathbf{J}_\Gamma^T)\, \mathrm{d}\mathbf{y}\right) \nabla_\mathbf{x} c_0\right) - \gamma|\partial \omega_f(\mathbf{x})|c_0, \quad (2.19)$$

where $\mathbf{I}$ is the 2×2 identity matrix. We now have a solvability equation for c_0, and can transform this into the leading-order homogenized equation for the intrinsic-averaged and volumetric-averaged concentrations.

2.3.2.4 The Homogenized Equations

We see from (2.4) that, at leading order, the intrinsic-averaged concentration is $\tilde{C}(\mathbf{x}, t) \sim c_0(\mathbf{x}, t)$ and the volumetric-averaged concentration is $C(\mathbf{x}, t) \sim \phi(\mathbf{x})c_0(\mathbf{x}, t)$. After a final rearrangement of (2.19), using $|\omega_f(\mathbf{x})| = \phi(\mathbf{x})$ from (2.3), converting functions of $\mathbf{x}$ into functions of $\phi(\mathbf{x})$ for convenience, and suppressing the argument for brevity, we obtain the homogenized equation for the intrinsic-averaged concentration

$$\frac{\partial \tilde{C}}{\partial t} = \frac{1}{\phi}\nabla_\mathbf{x} \cdot \left(\phi D(\phi) \nabla_\mathbf{x} \tilde{C}\right) - k(\phi)\tilde{C}. \quad (2.20)$$

Noting that $\phi = \phi(\mathbf{x})$, so we cannot bring factors of ϕ unchanged through a derivative, the homogenized equation for the volumetric-averaged concentration is

$$\frac{\partial C}{\partial t} = \nabla_{\mathbf{x}} \cdot \left[D(\phi) \left(\nabla_{\mathbf{x}} C - \frac{\nabla_{\mathbf{x}} \phi}{\phi} C \right) \right] - k(\phi) C, \qquad (2.21)$$

where the effective diffusion, $D(\phi)$ and the effective uptake, $k(\phi)$, are defined as

$$D(\phi) = 1 - \frac{1}{\phi} \int_{\omega_f(\phi)} \frac{\partial \Gamma_1}{\partial y_1} \, d\mathbf{y}, \qquad (2.22a)$$

$$k(\phi) = \gamma \frac{\sqrt{4\pi(1-\phi)}}{\phi}, \qquad (2.22b)$$

and $k(\phi)$ is obtained using $|\partial \omega_f| = 2\pi R = \sqrt{4\pi(1-\phi)}$.

The effective diffusion $D(\phi)$ is a scalar rather than a matrix (as it appears in (2.19)) because the cell problems for the components of $\boldsymbol{\Gamma}$, given in (2.13), are symmetric across both the y_1- and y_2-axes. Hence, $\mathbf{J}_{\boldsymbol{\Gamma}}$ is a multiple of the identity matrix and thus we could also use $\partial \Gamma_2 / \partial y_2$ as the integrand in (2.22a) instead of $\partial \Gamma_1 / \partial y_1$, with the same result. We also note that D is not strictly an effective diffusion coefficient in the homogenized equation for the intrinsic-averaged concentration (2.20) because of the factor of $1/\phi$ outside the derivative and the factor of ϕ inside the derivative. The effective diffusion can be computed by solving the cell problem (2.13) for a given cell porosity ϕ, which is determined by the radius of the circular obstacle. We do this numerically using the finite-element software COMSOL Multiphysics. The effective diffusion monotonically increases from 0 to 1 as the porosity increases from $1 - \pi/4 \approx 0.21$ to 1 or, equivalently, as the circle radius decreases from 0.5 to 0 (Fig. 2.2a). As the microstructure gets closer to blocking and ϕ gets closer to $1 - \pi/4$, the diffusion coefficient sharply decreases towards 0. This is because it is very difficult for a drug particle in one region to diffuse to another region when the gaps between nearly touching circles get very small.

The effective uptake $k(\phi)$ requires no further problems to be solved, as an explicit representation is given in (2.22b). The effective uptake monotonically decreases from $\pi/\sqrt{1 - \pi/4} \approx 14.64$ to 0 as the porosity increases from $1 - \pi/4 \approx 0.21$ to 1 or, equivalently, as the circle radius decreases from 0.5 to 0 (Fig. 2.2b). The effective uptake vanishes when the porosity approaches 1 because, in this limit, there is no tissue available to absorb the drug.

2.4 Interpreting the Homogenized Problem

The two homogenized equations (2.20) and (2.21) are equivalent through the transformation $\tilde{C}(\mathbf{x}, t) = \phi(\mathbf{x}) C(\mathbf{x}, t)$. We discuss the implications of the homogenized problem by referring only to (2.21), the homogenized equation for the

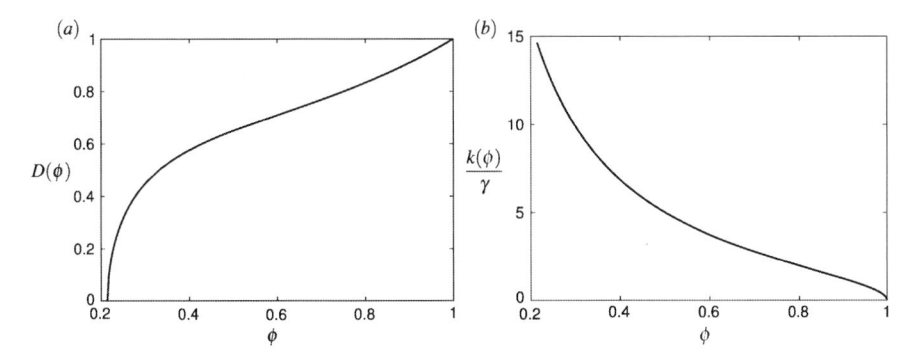

Fig. 2.2 The effective diffusion and uptake coefficients for $\phi \in (1 - \pi/4, 1)$. (**a**) The effective diffusion $D(\phi)$. (**b**) The (scaled) effective uptake $k(\phi)/\gamma$. As $\phi \to (1 - \pi/4)^+$, the domain tends to that of touching circles where the fluid domain becomes disconnected, and thus the effective diffusion vanishes. As $\phi \to 1^-$, the circular obstacles have a vanishingly small radius. This limit is regular, and thus the effective diffusion and uptake attain the values one would expect. That is, the effective diffusion is 1, and the effective uptake vanishes

volumetric-averaged concentration, as this equation is in standard advection–diffusion–reaction form.

We see that the reaction term, which arises from the Robin boundary condition modelling uptake on the boundary in the full problem, now appears as a bulk uptake, weighted by the boundary length to fluid phase area in a cell. This bulk term is one that we may have anticipated from the start.

Perhaps more surprisingly, the macroscale variation of the porosity has resulted in the presence of an advection term in the direction of a negative porosity gradient, as identified in [3, 7, 8]. This term advects the dissolved drug towards regions of larger porosity, and can be understood by noting that (2.21), the homogenized equation, is an equation for the concentration averaged over the *entirety* of one cell, including the tissue phase. Thus, as the area of the fluid phase is greater within cells with a larger porosity, the intrinsic smoothing of diffusion will act to increase the concentration averaged over an entire cell in cells with a larger porosity.

We additionally note that when the cell-averaged porosity is constant, and there is no macroscale variation in the microscale, the homogenized equations for the intrinsic-averaged and volumetric-averaged concentrations (2.20)–(2.21) reduce to the result that could be obtained from a traditional homogenization procedure with a strictly periodic microscale.

Appendix 1

When carrying out a homogenization using the method of multiple scales, the variation of the macroscale variable x across one cell is often ignored, and x is treated as a constant when the microscale variable varies (this microscale variable can be formally defined as $y = x/\epsilon - \lfloor x/\epsilon \rfloor$). Note that we just consider a one-dimensional problem in this appendix, but the argument we present generalizes to higher dimensions.

If there is a boundary condition or a boundary within a cell that depends on x and y, this should vary slowly with x through a cell. In this section, we see that it is fine to treat x as a constant when prescribing a boundary or for a Robin boundary condition. Note that this is *not* true for integral constraints, and the reasons for this are investigated in [4]. We first show that it is fine to treat x as a constant within one cell when working with a boundary that depends on x and y, then show that the Robin boundary condition is also unchanged.

Consider a problem where the boundary is defined as

$$f(x, y) = 0 \tag{2.23}$$

within a cell. Then, using the asymptotic expansion $f = f_0 + \epsilon f_1 + \cdots$, writing $x = x_0 + \epsilon y$, where $x_0 = \epsilon \lfloor x/\epsilon \rfloor$, and expanding the first argument in a Taylor series, we obtain

$$f(x_0 + \epsilon y, y) = \sum_{m=0}^{\infty} \epsilon^m \sum_{n=0}^{m} y^n \frac{\partial^n f_{m-n}}{\partial x^n}(x_0, y) = 0. \tag{2.24}$$

Note that we can change x_0 by an $O(\epsilon)$ amount (that is, evaluate x_0 anywhere within a cell), and (2.24) remains unchanged when evaluating $f(x, y)$.

We proceed by showing $f_i(x_0, y) = 0$ for all $i \geq 0$ by induction, and hence that the $O(\epsilon^i)$ perturbation to the interface is the same if we perturb x by $O(\epsilon)$. Namely, that it does not matter where we evaluate the macroscale variable within a cell.

The result for $i = 0$ follows from the $O(1)$ equation. The $O(\epsilon^I)$ equation is

$$\sum_{n=0}^{I} y^n \frac{\partial^n f_{I-n}}{\partial x^n}(x_0, y) = 0. \tag{2.25}$$

Under the induction assumption, that $f_i(x_0, y) = 0$ for all integers i such that $0 \leq i \leq I - 1$, all terms apart from $n = 0$ vanish immediately, yielding

$$f_I(x_0, y) = 0, \tag{2.26}$$

as required, thus showing that $f_i(x_0, y) = 0$ for all integers $i \geq$, and that we can evaluate the macroscale variable anywhere within a cell when defining a boundary.

We now consider a general Robin boundary condition

$$u'(x, y) + \alpha u(x, y) + \beta = 0. \tag{2.27}$$

Under the multiple scales transformation, where $\partial_x \mapsto \partial_x + \epsilon^{-1} \partial_y$, using the asymptotic expansion $u = u_0 + \epsilon u_1 + \cdots$, and writing $u(x, y) = u(x_0 + \epsilon y, y)$, the Robin boundary condition (2.27) becomes

$$\sum_{m=0}^{\infty} \epsilon^m \sum_{n=0}^{m} y^n \left(\frac{\partial^{n+1} u_{m-n}}{\partial y \partial x^n} + \epsilon \left(\frac{\partial^{n+1} u_{m-n}}{\partial x^{n+1}} + \alpha \frac{\partial^n u_{m-n}}{\partial x^n} \right) \right) + \epsilon \beta = 0, \qquad (2.28)$$

evaluated at (x_0, y).

Our induction hypothesis is that the standard multiple scales transformation and asymptotic expansion for u will be valid no matter where we evaluate x within a cell. That is, we wish to show that $\partial_y u_i(x_0, y) + \partial_x u_{i-1}(x_0, y) + \alpha u_{i-1}(x_0, y) + \beta \delta_{1i} = 0$ for all integers $i \geq 0$, where $u_j \equiv 0$ for $j < 0$, and δ_{kl} is the Kronecker delta.

The result for $i = 0$ follows from the $O(1)$ equation. The $O(\epsilon^I)$ equation is

$$\sum_{n=0}^{I} y^n \frac{\partial^n}{\partial x^n} \left(\frac{\partial u_{I-n}}{\partial y} + \frac{\partial u_{I-n-1}}{\partial x} + \alpha u_{I-n-1} \right) + \beta \delta_{1I} = 0, \qquad (2.29)$$

evaluated at (x_0, y). Under the induction hypothesis, $\partial_y u_i + \partial_x u_{i-1} + \alpha u_{i-1} + \beta \delta_{1i} = 0$ at (x_0, y) for all integers i such that $0 \leq i \leq I - 1$, all terms inside the sum apart from the $n = 0$ term vanish, leaving

$$\frac{\partial u_I}{\partial y} + \frac{\partial u_{I-1}}{\partial x} + \alpha u_{I-1} + \beta \delta_{1I} = 0, \qquad (2.30)$$

evaluated at (x_0, y), as required. Thus, we have shown that $\partial_y u_i + \partial_x u_{i-1} + \alpha u_{i-1} + \beta \delta_{1i} = 0$ at (x_0, y) for all integers $i \geq 0$, and that we can evaluate the macroscale variable anywhere within a cell when using a Robin boundary condition.

Finally, note that we have written $R(\mathbf{x})$ in (2.5) as a continuous function of the macroscale variable $\mathbf{x}$, rather than a piecewise constant function evaluated at the centre of the relevant unit cell. This simplifies the subsequent analysis while affecting only the boundary condition (2.7b) at higher orders than we need to consider. As a result, our final leading-order macroscale equation (2.21) is unchanged by employing this simplification.

Appendix 2

The transport theorem allows one to differentiate through an integral where the domain of integration depends on the variable of differentiation. In general, this yields two different terms: one from the variation of the integrand over the domain, and another from the variation of the domain with respect to the variable of integration. The transport theorem is often used to calculate derivatives with respect to time, and hence the second term often involves the velocity of the boundary. However, we are interested in the derivative with respect to space. In particular, with respect to the macroscale variable $\mathbf{x}$.

We note that any variation between two different fluid regions $\omega_f(\mathbf{x}_1)$ and $\omega_f(\mathbf{x}_2)$ is due to the difference in the radius of the circle within each cell, given by $R(\mathbf{x}_1)$ and $R(\mathbf{x}_2)$, respectively, and is not affected by the outer boundary, given by $\omega(\mathbf{x}_1)$ and $\omega(\mathbf{x}_2)$, respectively. Moreover, the rate of change of the circle $\omega_s(\mathbf{x}) = \omega(\mathbf{x}) \setminus \omega_f(\mathbf{x})$ with respect to $\mathbf{x}$ is $\nabla_\mathbf{x} R$. This can be deduced by considering the difference over the domains $\omega_f(\mathbf{x})$ and $\omega_f(\mathbf{x} + \xi \mathbf{e}_i)$ as $\xi \to 0$, where $\mathbf{e}_i$ is the unit vector in the direction of increasing x_i. The resulting domain of integration is a shell whose thickness is approximately $\xi \partial R / \partial x_i$ as $\xi \to 0$. As we are considering an integral over $\omega_f(\mathbf{x}) = \omega(\mathbf{x}) \setminus \omega_s(\mathbf{x})$, the relevant velocity of the interior boundary is $-\nabla_\mathbf{x} R$.

Therefore, for any vector function $\mathbf{v}(\mathbf{x}, \mathbf{y}, t)$, the relevant transport theorem is given by

$$\nabla_\mathbf{x} \cdot \int_{\omega_f(\mathbf{x})} \mathbf{v} \, \mathrm{d}\mathbf{y} = \int_{\omega_f(\mathbf{x})} \nabla_\mathbf{x} \cdot \mathbf{v} \, \mathrm{d}\mathbf{y} - \oint_{\partial \omega_f(\mathbf{x})} \nabla_\mathbf{x} R \cdot \mathbf{v} \, \mathrm{d}s. \tag{2.31}$$

Here, the first term on the right-hand side of (2.31) arises from the variation of the integrand over the domain and is relatively straightforward. The second term on the right-hand side of (2.31) arises from the variation of $\omega_f(\mathbf{x})$ with respect to $\mathbf{x}$, as described in the paragraph above.

Acknowledgements The author conceived the ideas for this book chapter during a previous appointment at the University of Oxford, where he was supported by an EPSRC Impact Acceleration Account Award (grant number EP/K503769/1). This work was also supported by the 2020 Science programme, which was funded through the EPSRC Cross-Disciplinary Interface Programme (grant number EP/I017909/1). Both of these projects were supervised by Maria Bruna and Ian M Griffiths. The author would like to thank Maria Bruna, S Jonathan Chapman, and Ian M Griffiths for helpful discussions concerning the content of this work.

References

1. Belyaev AG, Pyatnitskii AL, Chechkin GA (1998) Asymptotic behavior of a solution to a boundary value problem in a perforated domain with oscillating boundary. Sib Math J 39(4):621–644. https://doi.org/10.1007/BF02673049
2. Bensoussan A, Lions JL, Papanicolaou G (1978) Asymptotic analysis for periodic structures. North-Holland Publishing, Amsterdam
3. Bruna M, Chapman SJ (2015) Diffusion in spatially varying porous media. SIAM J Appl Math 75(4):1648–1674. https://doi.org/10.1137/141001834
4. Chapman SJ, Mcburnie SE (2015) Integral constraints in multiple-scales problems. Eur J Appl Math 26(05):595–614. https://doi.org/10.1017/S0956792514000412
5. Chechkin GA, Piatnitski AL (1999) Homogenization of boundary-value problem in a locally periodic perforated domain. Appl Anal 71(1–4):215–235. https://doi.org/10.1080/00036819908840714
6. Chernyavsky IL, Leach L, Dryden IL, Jensen OE (2011) Transport in the placenta: homogenizing haemodynamics in a disordered medium. Phil Trans Royal Soc A Math Phys Eng Sci 369(1954):4162–4182. https://doi.org/10.1098/rsta.2011.0170

7. Dalwadi MP, Griffiths IM, Bruna M (2015) Understanding how porosity gradients can make a better filter using homogenization theory. Proc R Soc A 471(2182). https://doi.org/10.1098/rspa.2015.0464
8. Dalwadi MP, Bruna M, Griffiths IM (2016) A multiscale method to calculate filter blockage. J Fluid Mech 809:264–289. https://doi.org/10.1017/jfm.2016.656
9. Fatima T, Arab N, Zemskov EP, Muntean A (2011) Homogenization of a reaction–diffusion system modeling sulfate corrosion of concrete in locally periodic perforated domains. J Eng Math 69(2–3):261–276. https://doi.org/10.1007/s10665-010-9396-6
10. Hornung U (2012) Homogenization and porous media, vol 6. Springer Science & Business Media, New York
11. Mei CC, Vernescu B (2010) Homogenization methods for multiscale mechanics. World Scientific, Singapore
12. van Noorden TL (2009) Crystal precipitation and dissolution in a porous medium: effective equations and numerical experiments. Multiscale Model Simul 7(3):1220–1236. https://doi.org/10.1137/080722096
13. van Noorden TL, Muntean A (2011) Homogenisation of a locally periodic medium with areas of low and high diffusivity. Eur J Appl Math 22(05):493–516. https://doi.org/10.1017/S0956792511000209
14. Peter MA, Böhm M (2009) Multiscale modelling of chemical degradation mechanisms in porous media with evolving microstructure. Multiscale Model Simul 7(4):1643–1668. https://doi.org/10.1137/070706410
15. Ray N, van Noorden T, Frank F, Knabner P (2012) Multiscale modeling of colloid and fluid dynamics in porous media including an evolving microstructure. Transp Porous Media 95(3):669–696. https://doi.org/10.1007/s11242-012-0068-z
16. Ray N, Elbinger T, Knabner P (2015) Upscaling the flow and transport in an evolving porous medium with general interaction potentials. SIAM J Appl Math 75(5):2170–2192. https://doi.org/10.1137/140990292
17. Richardson G, Chapman SJ (2011) Derivation of the bidomain equations for a beating heart with a general microstructure. SIAM J Appl Math 71(3):657–675. https://doi.org/10.1137/090777165
18. Sánchez-Palencia E (1980) Non-homogeneous media and vibration theory. In: Lecture notes in physics, vol 127. Springer, Berlin
19. Whitaker S (1998) The method of volume averaging, vol 13. Springer Science & Business Media, New York

Chapter 3
Mathematical Models for Acid-Mediated Tumor Invasion: From Deterministic to Stochastic Approaches

Sandesh Athni Hiremath and Christina Surulescu

3.1 Introduction

Tumors feature high metabolic rates by excess of glycolysis both under hypoxic and normoxic conditions [64, 69]. One of the prominent characteristics of cancer is its invasiveness, which is tightly connected to the upregulated glucose metabolism and essential to proteolytic tissue degradation and establishment of metastases at distant sites [29]. Indeed, there is wide evidence that reinforced glycolysis, a reversed pH-gradient (intracellular pH higher than extracellular pH), and acid resistance subsequently developed by tumor cells are conferring the latter a substantial growth advantage, boosting both proliferation and invasion [24, 62, 72]. The acidification of the peritumoral environment induces apoptosis of normal cells, thus allowing the tumor to extend into the space becoming available. Thus, hypoxia and acidity are believed to contribute—among other factors—to the progression from benign to malignant growth [23, 72] and hence to exert direct influence on the metastatic potential of the tumor cells [1, 42].

During the last two decades diagnostic methods relying on positron emission tomography (PET) and acidity quantification via glucose analogue tracers (e.g., fluorodeoxyglucose FDG, [38, 46, 74]) in the proximity of neoplastic lesions have become standard. This attracted increased interest in tumor metabolism and particularly in biological processes closely related to glycolytic metabolism, resulting in a large variety of therapeutic approaches [20, 28].

Mathematical modeling can provide a valuable tool for a deeper understanding of the complex phenomena related to acid-mediated tumor invasion and development.

S.A. Hiremath (✉) · C. Surulescu
Felix-Klein-Center for Mathematics, TU Kaiserslautern, Paul-Ehrlich-Str. 31, 67663 Kaiserslautern, Germany
e-mail: hiremath@mathematik.uni-kl.de; surulescu@mathematik.uni-kl.de

Moreover, mathematical models allow in silico testing of hypotheses, therapy approaches in various combinations, evaluation of treatment plans, interconnection of scales, the ad libitum activation and deactivation of mechanisms, all of which would be very expensive (if at all possible) in the wet lab. Their predictive power can facilitate the assessment of tumor growth and extent relying on adequate data which are either already existing or yet to be made available by the advancement of medical imaging and further diagnostic methods. This in turn is expected to enhance both therapy planning and therapeutic outcome.

A large variety of mathematical models describing manifold aspects of tumor development and invasion (e.g., cell population growth, tactic behavior, tissue degradation, treatment response, proton exchange through membrane transporters and proton buffering), with the involved processes conditioned by intra and extracellular pH have been proposed and studied. Depending on their respective approaches, they can be broadly classified into discrete and continuum models, with an intermediate class of so-called hybrid or semidiscrete settings accounting for continuum dynamics of soluble or insoluble components present in the microlocal tumor environment while handling the cancer cells individually. The (semi)discrete model class includes agent-based formulations and cellular automata which are able to encompass a rather detailed description of the biochemical and biophysical events in connection with the cell behavior, see e.g., [2] for discrete and [25, 50, 57] for hybrid approaches, respectively. While appealing in their simplicity, these discrete models (and also their semidiscrete versions) exhibit a certain degree of arbitrariness in their choice of rules and due to the algorithms used to solve for the development and dispersal of cells giving different outputs for the same initial condition; hence these models can become very expensive, especially for large cell populations and high levels of detail in the description. Moreover, due to the lack of a well established mathematical framework there are no guidelines for the number of runs necessary for a valid prediction within a given degree of certainty. These drawbacks are avoided by using continuum approaches for which numerical methods with rigorous error analyses are (to a large extent) well developed. Therefore, we shall only address here continuum settings, although their discrete counterparts can have some other advantages, as they account for details which cannot be considered in a continuum setting.

This chapter is organized as follows: Sect. 3.2 gives a review of models addressing issues connected to hypoxia, acidosis, acid-mediated invasion, or proton dynamics. Thereby, the main part consists of deterministic models, most of them acting on a single scale (either subcellular or macroscopic). Some recent multiscale approaches are recalled as well. The stochastic models are far more scarce than their deterministic counterparts; in Sect. 3.3 we explicitly re-address our recent stochastic multiscale models [31, 32]. Finally, in Sect. 3.4 we provide a discussion about perspectives of mathematical models for acid-mediated tumor invasion, thereby paying attention both to the mathematical issues and their applicability potential.

3.2 Continuum Mathematical Models: A Synopsis

3.2.1 *Deterministic Approaches*

Presumably the first continuum mathematical model for acid-mediated tumor invasion was introduced by Gatenby and Gawlinski [22] and involved a system of reaction-diffusion equations characterizing the evolution of tumor and normal cell densities interacting with the dynamics of proton concentration. An elementary stability analysis hinted at a traveling wave whose speed could be determined and was found to be in agreement with experimental data. Furthermore, in a certain parameter regime the model predicted the existence of an interstitial gap at the interface between tumor and stroma; such very sparsely populated, that far not observed regions were subsequently identified in a series of experiments, in which this phenomenon occurred on a non-regular basis (in 14 out of 21 specimens). Although the well-posedness of that model was not investigated there, in spite of the presence of degenerate diffusion for both tumor and normal cells which arises some nontrivial mathematical questions, it inspired a plethora of models. A more detailed analysis of the properties of traveling wave associated with a one-dimensional version of the model in [22] has been done in [18], also computing the width of the interstitial gap stipulated therein, however without rigorously proving the existence of traveling waves. Similar analyses of traveling wave properties for versions of this model involving: nonlinear acid production, tumor depletion by excessive acidity, inter- and intraspecies competition, or accounting for proton buffering have been subsequently performed in [33, 43, 44].

Further related models assuming radially symmetric tumor expansion and investigating the role of acidosis in the interaction between normal and tumor cell populations in the vascular and/or avascular cases—some also accounting for cell quiescence—have been proposed and analyzed e.g., in [8, 56, 58]. These 1D models are related to more complex reaction-diffusion-taxis (RDT) settings characterizing the dynamics of solid (vascular or avascular) tumors interacting with the extracellullar matrix (ECM), which they can degrade by proteolysis, some evolving in the presence of angiogenesis, which is tightly connected to hypoxia and any tumor acidity issue, see e.g., [7, 15, 29]. Thereby, the tumor cells perform chemotaxis (i.e., they follow the concentration gradient of some soluble and diffusive chemoattractant), haptotaxis (cell orientation towards the gradient of some spatially fixed stimulus, e.g., density of ECM fibers), or both types of taxis. Quite a few variants of such models have been studied (we mention here only [10] and its many variants subsequently analyzed, among others, by Tao and Winkler e.g., in [67, 68]). However, RDT models explicitly accounting for acid-mediated invasion and associated proliferation events are rare; we are actually aware of only the multiscale settings in [45, 60], both concerned with the so-called pH-taxis. The latter refers to the tumor cells biasing their motion in response to the pH-gradient in their surroundings, as proposed in [4, 49]. The large reaction-diffusion model in [65] describes glioma invasion in the presence of angiogenesis, thereby differentiating between normoxic, hypoxic, and necrotic cancer cells, however completely neglecting any kind of taxis and allowing for degenerate diffusion. The latter poses very challenging analytical and numerical issues e.g., when haptotaxis (a highly relevant issue in that context, due to the anisotropy of brain tissue, see e.g. [16]) is accounted for.

The above models are—with exception of those in [45, 60]—purely macroscopic: They characterize the space-time dynamics of cell densities and chemical concentrations on the population level. Another class of models related to tumor acidity describe biochemical events taking place on the subcellular scale. In particular, such models investigate the mutual dependence between the activity of several membrane ion transport systems and the changes in the peritumoral space [3, 70, 71]. These settings also involve intracellular proton buffering, effects of the expression/activation of matrix degrading enzymes (MDEs), and proton removal by vasculature. Moreover, [70] also accounts for the influence of alkaline intracellular pH on the growth of tumor cells, which makes the connection to the macroscopic level and thus can be seen as a first step towards multiscale modeling. Genuinely multiscale models, however, involve explicitly at least two scales, usually the subcellular and the macroscopic one. Since migration and invasiveness, which have pronounced spatial effects (observable both on the cellular and the population level), are crucially influenced by subcellular dynamics, a higher dimensional and more complex modeling framework is required, which can couple the subcellular level with the mesoscale (accounting for cell motion directions) and/or the macrolevel dynamics. Depending on these couplings, the resulting models are sometimes called micro-meso, micro-macro, or micro-meso-macro models.

The micro-macro model in [60] features on the microscopic (subcellular) scale the dynamics of intracellular H^+ concentration and its exchange with the extracellular protons; the concentration of the latter is a macroscopic quantity satisfying a reaction-diffusion equation (RDE). It acts as guiding agent for the pH-taxis performed by the tumor cells supplementary to their diffusive spread. This RDT-ODE-RDE system is completed by an ODE for the evolution of normal tissue being degraded by the extracellular protons and restructured in concurrence with the tumor cells:

$$\begin{cases} \partial_t c = \nabla \cdot (\varphi(c,n)\nabla c) - \nabla \cdot (f(h,c)\nabla h) + \mu_c(y)c\left(1 - \frac{c}{K_c(h(\cdot,t-\tau))} - \eta_1 \frac{n}{K_n}\right) \\ \partial_t n = -\delta_n hn + \mu_n n\left(1 - \eta_2 \frac{c}{K_c(h(\cdot,t))} - \frac{n}{K_n}\right) \\ \partial_t h = D_h \Delta h + \Theta(y,h) \\ \partial_t y = -\Theta(y,h) - \alpha y + g(c) \end{cases}$$

(3.1)

Here, c and n denote the macroscopic densities of tumor cells and normal tissue, respectively, while h and y are the extra- and intracellular proton concentrations, respectively. The nonlinear coefficients $\varphi(c,n)$ for diffusion and $f(h,c)$ for pH-taxis model the dependence of motility on the surrounding acidity and the adaptation of motion to the amount of neighboring cells. The proliferation rate depends on the cytosol alkalinization and the function $\Theta(y, h)$ models the proton exchange over the cellular membranes. The carrying capacity of the neoplastic tissue depends on the surrounding acidity (expressed by the concentration of extracellular H^+) and is time-delayed (τ denotes the delay), in order to account for its non-instantaneous adaptation to acidosis. The global well-posedness of this system (supplemented by no-flux boundary conditions and adequate initial data) was proved, but the global boundedness of the weak solution remains open, unless some additional bounds are assumed for the coefficient functions φ and f. The performed numerical simulations showed that choosing a time-dependent carrying capacity for the tumor allows the latter to dynamically adapt its growth to the environmental acidity, although it does not seem to have a notable influence on the invasion speed. This endorses the idea of reducing the acidity in the tumor region in order to control its development, see e.g., [23, 66].

Still in the context of therapeutic outcome, it is well known that tumors are heterogeneous, more or less compact structures made up of cells featuring different phenotypes, according e.g., to their position in the cell cycle (active vs. quiescent), their migratory behavior (actively migrating vs. resting and proliferating), their sensitivity towards therapies etc. These classifications are by no means exhaustive, neither are the categories disjoint. Following the idea of tumor heterogeneity with respect to the local levels of hypoxia and its effects on proliferation, cell survival, and tumor recurrence we proposed in [45] a micro-macro model for the dynamics of a tumor composed of two subpopulations of cancer cells, one of which is actively proliferating and the other inferring a migratory phenotype, which goes along with suppressed proliferation. This differentiation between migrating and proliferating cells conforms with the go-or-grow dichotomy [13, 21, 26, 73]. Both migration and proliferation are crucially influenced by the gradients between intra- and extracellular pH and nutrient deprivation due to hypoxia, see e.g., [27, 30]. Moreover, therapeutic agents are more effective against cells in their mitotic phase and the therapy outcome depends dramatically on the tumor environmental pH [14, 19, 40, 59]. Motivated by these biological facts we associated in [45] the proliferating cells to the non-migrating and sensitive[1] phenotypes and assigned to the other subpopulation of tumor cells the motile, non-proliferating, and less sensitive ones. The following PDE-ODE system characterizes the evolution of the two subpopulations in mutual interaction ($a(x, t)$: density of sensitive, proliferating cells, $q(x, t)$: density of less sensitive, motile cells), their effect on the normal tissue n, and the influence of extra- and intracellular protons ($h(x, t)$ and $y(t)$, respectively);

---

[1]The sensitivity is understood with respect to (chemo and/or radio)therapy: cells in the mitotic phase are known to have a more pronounced response towards drugs and ionizing radiation, therefore we use the go-or-grow dichotomy to associate the proliferative—hence non-motile— phenotype with the more sensitive cells.

the quantity m represents the concentration of a cytotoxic agent acting on both tumor cells and normal tissue:

$$
\begin{cases}
\partial_t a = \underbrace{\mu_c(y)a\left(1 - \frac{a+q}{K_c(h(\cdot, t-\tau))} - \eta_1 \frac{n}{K_n}\right)}_{\text{proliferation}} + \gamma(d_c)q - \lambda a - a\Gamma_a R_a(\alpha_a, d_r) \\
\qquad\quad - a\kappa_a(d_c)C_a(\theta_a, m), \\
\partial_t q = \underbrace{\nabla \cdot (\varphi(a, q, n)\nabla q)}_{\text{nonlinear diffusion}} \underbrace{-\nabla \cdot (f(a, q, h)\nabla h)}_{\text{pH-taxis}} + \lambda a - \gamma(d_c)q - q\Gamma_q R_q(\alpha_q, d_r) \\
\qquad\quad - q\kappa_q(d_c)C_q(\theta_q, m), \\
\partial_t n = -\delta_n hn + \mu_n n\left(1 - \frac{a+q}{K_c(h(\cdot, t))} - \frac{n}{K_n}\right) - n\Gamma_n R_n(\alpha_n, d_r) - n\kappa_n C_n(\theta_n, m), \\
\partial_t h = D_h \Delta h + \Theta(y, h) - \sigma_c d_c h \\
\partial_t y = -\Theta(y, h) - \alpha y + g(a+q), \\
\partial_t m = -\rho m + v_m(t),
\end{cases}
$$

(3.2)

where the terms of the form $j\Gamma_j R_j(\alpha_j, d_r)$, with $j \in \{a, q, n\}$, model effects of applied radiotherapy with doses d_r (in Gy) and tissue sensitivity α_j^2; the terms $j\kappa_j C_j(\theta_j, m)$ ($j \in \{a, q, n\}$) model the influence of chemotherapeutic agents, the function $\Theta(y, h)$ characterizes the exchange of protons across cell membranes, and $\gamma(d_c), \lambda \geq 0$ are transition rates between a-cells and q-cells; d_c denotes the doses of chemotherapeutic drug and $g(a+q)$ describes the intracellular proton production. Proving the global well-posedness for a model of this type is challenging even without haptotaxis and with nondegenerate diffusion, due inter alia to the switching between the two cell subpopulations: the moving cells act on the one side as source for the proliferating ones, and on the other side as decay term for themselves. Moreover, the lack of diffusion in the a-equation leads to less regular solutions, whereas uniqueness is only available under higher regularity assumptions on the data. For more details we refer to [45] and to [61], where the global existence of weak solutions to a related, but different, even more challenging model has been proved and for which the problem of boundedness and uniqueness of solutions is completely open.

3.2.2 Stochastic Approaches

While the vast majority of the available mathematical models for tumor invasion are deterministic, there is an obvious need to include stochasticity, as it is a relevant feature inherent to many biological processes occurring on all modeling levels. In

---

[2]More precisely, α_j represents lethal lesions inflicted on the population j by a single radiation track in a linear-quadratic (LQ) description (see e.g., [54]).

particular, it has been found to have a significant influence on subcellular dynamics and individual cell behavior, see e.g. [63]. This is not different when it comes to pH dynamics: experimental evidence shows that although all cells follow the same biochemical mechanisms, there are variations and uncertainties in the behavior of every single cell (e.g., due to a random environment, to intracellular variations in the expression of some proteins, and/or to differences between cells in a population). Moreover, the distribution of intracellular pH at any value of extracellular pH was found to be broader than what was predicted by theoretical models based on machine noise and stochastic variation in the activity of membrane-based mechanisms regulating intracellular acidity [39]. In addition, excess current fluctuations have been observed in the gating of the ion channels [34]. Thus, the noise involved is both of extrinsic and of intrinsic nature. There are relatively numerous works dedicated to deducing macroscopic RDT equations upon starting from stochastic differential equations (SDEs) (see e.g., [47, 48] to mention just the best known ones). Thereby, although the individual cell dynamics is stochastic, the equations obtained on the population scale are—with few exceptions, see e.g. [12]—deterministic. There are comparatively very few continuum stochastic models (mainly involving stochastic or random PDEs) featuring taxis and even lesser models have a multiscale character. Recently, we proposed and analyzed in [31, 32, 37] some continuum multiscale models related to acid-mediated tumor invasion. The setting in [37] couples a PDE for extracellular proton concentration (macroscale) with an SDE for intracellular protons (subcellular scale). Thereby, both local and nonlocal sample dependence versions have been considered. The models in [31, 32] feature couplings of PDEs with random ODEs (RODEs), the latter model also accounting for pH-taxis, cross diffusion, and chemorepulsion. We outline more details about these two PDE-RODE systems in the following section.

3.3 Multiscale Stochastic Models for Acid-Mediated Tumor Invasion

In this section we discuss two stochastic models, each for a specific aspect of acid mediated invasion. The first model (we call it shortly SMAMCI: *stochastic multiscale acid-mediated cancer invasion* model) is a rather general description of the invasive behavior of cancer, while the second one (shortly called SAIG: *stochastic acid invasion with gaps* model) is designed to reproduce the occurrence of gaps and infiltrative patterns during cancer progression. Both settings have a multiscale character, as they connect the subcellular level (dynamics of intracellular proton concentration) with the macroscale, on which cell proliferation and tissue encroachment are modulated by intracellular and extracellular acidity. The models couple several types of equations for the following unknowns: (1) intracellular proton concentration, denoted by H_i, (2) extracellular proton concentration, denoted by H_e, (3) cancer cell density, denoted by C, and (4) normal cell density, denoted by N. Concretely, the models take the following forms:

SMAMCI model:

$$\partial_t H_i = -T_1(H_i, H_e) - T_2(H_i, H_e) + T_3(H_i) + S_1(v) - Q(H_i) + F(\xi_t, H_i) \quad (3.3)$$

$$\partial_t H_e = T_1(H_i, H_e) + T_2(H_i, H_e) - T_3(H_i) - S_2(v)H_e + D_1 \Delta H_e. \quad (3.4)$$

$$\partial_t C = \Lambda_1(H_i)C\big(1 - \eta_C C\big) - \Lambda_1^+(H_i)\eta_N CN + \nabla \cdot \big(D_2(C, N)\nabla C\big), \quad (3.5)$$

$$\partial_t N = -\Lambda_3(H_e)CN \quad (3.6)$$

SAIG model:

$$\partial_t H_i = J(C)\big(-T_1(H_i, H_e) - T_2(H_i, H_e) + T_3(H_i) + S_1(v) - Q(H_i) + F(\xi_t, H_i)\big) \quad (3.7)$$

$$\partial_t H_e = J(C)\big(T_1(H_i, H_e) + T_2(H_i, H_e) - T_3(H_i)\big) - S_2(v)H_e + D_1 \Delta H_e$$
$$+ \nabla \cdot \big(g(H_i, H_e, C)\nabla C\big) + \nabla \cdot \big(h(H_e, N)\nabla N\big) \quad (3.8)$$

$$\frac{\partial}{\partial t}C = C(1 - C)(\Lambda_1 + \Lambda_2) + \nabla \cdot (a\nabla C) - b\nabla H_e \cdot \nabla C \quad (3.9)$$

$$\partial_t N = -\gamma_N CN + N(1 - N)\big(-\Lambda_3 + \Lambda_4\big) \quad (3.10)$$

3.3.1 Description of the Models

3.3.1.1 Microscopic Dynamics: The Intracellular Proton Concentration

The dynamics of intracellular protons is given by the random differential equations (3.3) and (3.7) for the SMAMCI and the SAIG model, respectively. Thereby, T_1, T_2, and T_3 are real valued functions representing NDCBE (Na^+ dependent Cl^--HCO_3^- exchanger), NHE (Na^+ and H^+ exchanger), and AE (Cl^--HCO_3^- or *anion exchanger*) transporters, respectively.

The exact algebraic formulae for T_1 and T_2 were acquired by qualitatively reproducing the curves experimentally obtained (by Boyer and Tannock [6]) for the efflux of protons due NDCBE and NHE in MGH-U1 cell lines. However, for the T_3 function we followed [71] and made it a monotone decreasing function of H_i; this is motivated by the fact that AE acts as a counter-mechanism for the alkalinization of cytoplasm. Furthermore, Q denotes the function representing the loss of free protons due to intracellular buffering (e.g., by organelles). The function S_1 in (3.3) represents the observed constant acid production rate in cancer cells due to aerobic glycolysis. It is parameterized by the density of tissue vasculature v.

Finally, we use a random term to represent the ensemble of uncertainties influencing the proton concentration (e.g., by inter- and intracellular fluctuations, transporter fluctuations, and further stochastic effects). The noise is chosen to be a state dependent noise of the following form:

$$F(\xi_t, H_i) := F_1(H_i)F_2(\xi_t) := \vartheta H_i \xi_t, \qquad (3.11)$$

where for the SMAMCI model ξ_t is chosen to be a Brownian bridge process $B_t^{a,b}$ starting at $a \in \mathbb{R}$ and ending at $b \in \mathbb{R}$, while for the SAIG model ξ_t is taken to be an Ornstein-Uhlenbeck process O_t. However, the particular choice of the stochastic process ξ_t is not too important, other choices are equally conceivable.

For the SAIG model, the intracellular proton dynamics is modulated by a non-negative function J depending on the cancer cell density. It is constructed in a such way that the proton dynamics is dominant at the periphery of the tumor core.

3.3.1.2 Extracellular Proton Concentration

The next quantity of interest is the extracellular proton concentration $H_e(t, x)$. For the SMAMCI model the dynamics satisfy the reaction-diffusion equation (3.4), where the transport functions T_1, T_2 and T_3 are as mentioned above. The function S_2 is used to describe the removal of acid (protons) from the extracellular (interstitial) space by vasculature and takes the form $S_2(v) := a_5 \kappa v$. For the diffusion of extracellular protons diffuse in the interstitial space we consider a constant coefficient $D_1 > 0$.

More interestingly, the SAIG model includes two additional nonlinear operators, namely $\nabla \cdot (g \nabla C)$ and $\nabla \cdot (h \nabla N)$, in order to characterize the accumulation of acid at the tumor and stroma interface. The coefficient functions g and h are non-negative and real-valued, and serve to modulate the speed and direction of the flow of H_e. These operators phenomenologically capture the effect of proton repulsion from highly dense regions of cancer and normal cells towards less dense regions. This can be seen to be occurring due to fluid (acid) expulsion from a growing tumor and also due to the electrical field generated by the resting membrane potential, which is positive relatively to the intracellular side [11, 51], thus exerting a repulsive force on the positively charged protons. For more explanations we refer to [32].

3.3.1.3 Cell Dynamics on the Macroscale

In this section we present the equations for the evolution of the tumor cell density $C(t, x)$ and the normal cell density $N(t, x)$. For the SMAMCI model the dynamics of the cancer population density is described again by a nonlinear reaction diffusion equation (3.5).

The reaction term is given by a damped Lotka-Volterra function, which has a scaling prefactor depending on the intracellular proton concentration. The movement of cancer cells is modeled by a nonlinear diffusion term (represented in divergence form) with the diffusion coefficient D_2 being a function of cancer cell and normal cell densities.

The diffusion coefficient is inversely proportional to both quantities; this models the motion of cells through the crowded peritumoral region. The function $\Lambda_1^+(H_i)$ denotes the positive part $\max(\Lambda_1(H_i), 0)$ of $\Lambda_1(H_i)$.

The proliferation function $\Lambda_1(H_i)$ is used to capture the influence of intracellular proton concentration on the growth of cancer cells and has a form like in Fig. 3.1. For practical and natural values of H_i the function Λ_1 is positive, which for alkaline pH values models the enhancement of cancer proliferation. According to [55, 72], this may be due to resistance of cells to apoptosis, enhancement of cell division, and possibly even due to suppression of mitotic arrest [72]. Since even cancer cells are sensitive to extreme pH values, for such unrealistic values of pH (both in the alkaline and in the acidic directions) the proliferation function $\Lambda_1(H_i)$ takes negative values. The negative Λ_1 results in the decay of cancer cell density, which represents cell apoptosis due to e.g., activation of DNase II at low pH$_i$ (see [9]). Moreover, in [9] is also indicated a positive correlation between high pH$_i$ and cellular apoptosis.

In the SAIG model the cancer cell density satisfies an advection-reaction-diffusion equation (3.9). Here the reaction term is given by a Lotka-Voterra model modulated by intracellular proton dependent growth and decay functions Λ_1 and Λ_2, respectively. The nonlinear diffusion term depends on all quantities of interest: intra- and extracellular protons, cancer cell and normal cell densities. The dependence is specified in the diffusion coefficient $a > 0$, which takes values on a compact subset

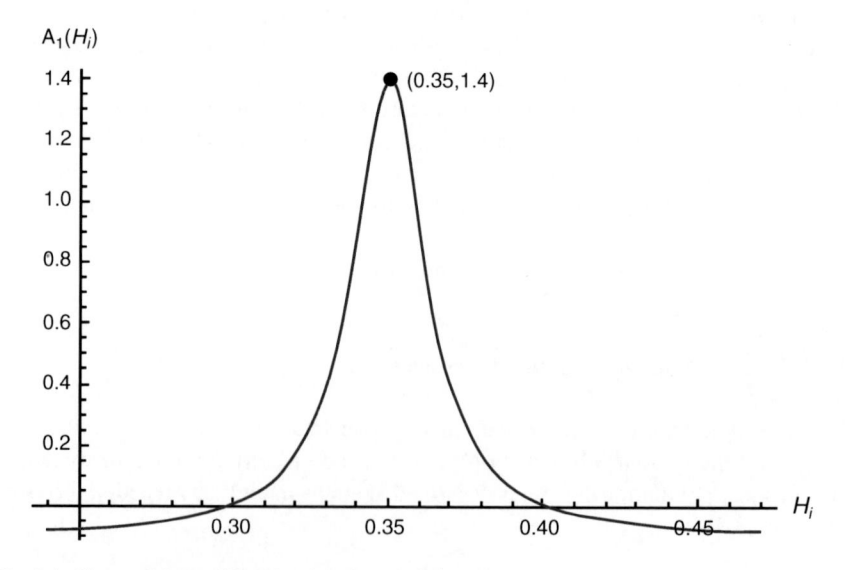

Fig. 3.1 Shape of the proliferation function $\Lambda_1(H_i)$ (see [31])

of the real line. The advection term models *pH*-taxis, i.e., motion in the direction of increasing acidity[5, 36, 53, 72]. The directed motion of cells and the corresponding velocity are governed by b, a nonlinear function of H_i and H_e. An interesting aspect is that the functions Λ_1, Λ_2 and b representing growth, decay, and directed motion, respectively, have mutually exclusive support, i.e. for fixed time and spatial point only one of the three function is larger than 0. This is in accordance with the *go or grow or decay* hypothesis of cancer cells being either in motion, proliferating, or depleting, none of these options occurring simultaneously with another [13, 21].

Finally, we describe the evolution of the normal cell density, which is supposed to be degraded by the invading tumor. For the corresponding equation (3.6) in the SMAMCI model there is only a decay term ignoring remodeling. Thereby, the depletion amplifying factor Λ_3 depends on the concentration of extracellular protons.

In Eq.(3.10) of the SAIG model the cell regression is due to direct interaction with cancer cells and to the degradation by extracellular acidity; the latter is modeled by the function $\Lambda_3(H_e)$. The logistic growth is modulated by the function Λ_4 depending on H_e and capturing e.g., the immune response induced by extracellular acidity.

3.3.2 Analytical Results

Due to the nonlinear coupling and the possibly unbounded (with respect to time) stochastic process, it is important to know what conditions must be imposed on the involved coefficients and on the initial condition in order to be able to solve the equation for a unique non-negative solution. This calls for a well-posedness analysis of the stochastic models (SMAMCI and SAIG) presented above.

For the SMAMCI model we use the weak formulation of the problem and look for a unique weak solution.[3] Under biologically meaningful assumptions such as: uniformly bounded reaction terms, bounded domain $\mathfrak{D} \subset \mathbb{R}^n$, and some technical assumptions such as: Lipschitz continuity of reaction terms, regularity of the initial conditions, regularity of the domain boundary, semi-martingale property for the noise term, and restriction of the original probability space $(\Omega, \mathscr{A}, \mathbb{P})$ to a new, smaller probability space $(\Omega_\epsilon, \mathscr{A}_\epsilon, \mathbb{P}_\epsilon)$ containing those paths of the noise process that are bounded from above by an arbitrarily predetermined upper bound, we proved the existence of a unique non-negative weak solution. The obtained solution is an $L^2(\Omega_\epsilon)$ process taking values in a closed subspace of $C([0, T]; L^2(\mathfrak{D}))$ and is $\mathscr{A}_\epsilon$ adapted.

Theorem 3.1 ([31]) *For every $T < \infty$ there exists a unique weak solution to the SMAMCI model.*

---

[3]For the definition of a weak solution see [17] p. 379.

Proof (Sketch) The idea of the proof is to first establish some estimates for a slightly altered and approximated simpler problem, then use an iterative technique to generate a sequence of approximate solutions, and finally show that the sequence converges in an appropriate sense by using a Banach fixed point argument. For details of the proof we invite the reader to refer to Theorems 3.10, 3.13, and 3.20 to 3.23 in [31].

To verify the well-posedness of the SAIG model we resort to semigroup theory, since it directly allows us to treat the problem in $L^p(\mathfrak{D})$, $p > n$, $n \in \{1, 2, 3\}$. To do so, we reformulate the SAIG model as an abstract Cauchy problem for each $\omega \in \Omega$ and look, pathwise, for a unique mild solution.[4] Then we identify the spatial operators and their domains restricted to the $L^p(\mathfrak{D})$ space. In the equations for H_e and C the diffusion operators (Δ and $\nabla \cdot (a\nabla)$) are the semigroup generating operators whose domains (restricted to $L^p(\mathfrak{D})$) turn out to be Bessel potential spaces $H_p^2(\mathfrak{D})$ (see Definition 3.2 in Sect. 3.4 below). Since the equations for H_i and N are spatially uncoupled, they are treated as inhomogeneous ODE equations. The initial conditions are chosen to be in $H_p^k(\mathfrak{D})$ for appropriate $k \geq 1$ and $p \in \mathbb{N}$. The involved coefficient functions are assumed to satisfy some mild growth conditions and as functions of time they are assumed to be in some Banach algebra i.e., $H_p^k(\mathfrak{D})$. The latter allows for an easy verification of the Lipschitz property. Finally, the noise term is assumed to have Gaussian law with independent increments and a.s. continuous paths. Under these conditions, we get the following well-posedness result:

Theorem 3.2 ([32]) *For every $T < \infty$ there exists a unique mild solution to the SAIG model.*

Proof (Sketch) We first construct an evolution operator and then use it to generate a sequence of approximate mild solutions, each of which solves the respective approximated abstract Cauchy problem. Finally, the sequence is shown to be contractive, thereby enabling us to invoke the Banach fixed point theorem. This establishes the local existence (see Theorem 4.8 in [32]), which is then extended continuously to any finite time interval $[0, T] \subset [0, \infty)$. The solution exists as an $L^2(\Omega)$ process and takes values in an appropriate closed subset of $C_{\{0\}}^\mu([0, T], X)$, where X denotes an appropriate Bessel potential space $H_p^k(\mathfrak{D})$. Furthermore, it is also shown that the solution is unique and non-negative (see Lemmata 4.9 to 4.11 in [32]).

3.3.3 Simulation Results

Next we present and discuss some simulation results for both SMAMCI and SAIG models. We refrain here from giving any description on the choice of parameters

---

[4]For the definition of a mild solution refer to Definition 2.3 of Section 4.2 of [52], p. 106.

and the involved model functions and refer to [31, 32] for such details, focusing instead on the main insights.

3.3.3.1 Simulations for the SMAMCI Model

The expected behavior of the tumor invasion as predicted by the numerical simulation of the SMAMCI model is shown in Fig. 3.2. The expectation was numerically obtained by averaging over ~ 400 sample paths. The cancer cell density C behaves like a traveling wave with the cell population spreading over the neighboring region and invading the stroma N, while the normal cells show a gradual decay as the cancer wave advances. Also notice that over time the extracellular proton concentration H_e becomes higher than its intracellular counterpart H_i, thus confirming the expected feature of a reversed pH-gradient [72]. To get a more detailed insight into the respective dynamics we look at different sample paths of the solutions to the SMAMCI model. Figures 3.3 and 3.4 depict the time evolution plots of the involved quantities at different, uniformly spaced positions. From these figures we infer the following: (1) The SMAMCI model predicts a decay, with respect to time, of the cancer cell density $C(t, x)$. This is due to the positive to negative switching property of the proliferation function $\Lambda_1(H_i)$. Thus, depending on the dynamics of the intracellular protons, $\Lambda_1(H_i)$ can take values in the support of $\Lambda_1(H_i)^-$ (the negative part of the function Λ_1). Such extreme values of H_i represent cytotoxicity and are induced mainly by the noise term. Thus, consistent cytotoxic H_i values induce a sustained decay in the cancer cell density. This behavior can be seen in Fig. 3.3, where decay (even down to zero) occurs in some of the sample paths of $C(t, \cdot)$. (2) From Fig. 3.3 we see that the expected behavior of cancer is that of a slow but gradual increase in density (which represents slow invasion), although some of the sample paths of $C(t, x)$ exhibit consistent decay. As a result, on average the rate of invasion is significantly reduced. Furthermore, an interesting observation is that the smaller the proliferation sensitivity parameter in $\Lambda_1(H_i)$, the higher is the expected cancer cell density. This can be seen by comparing Fig. 3.3 with Fig. 3.4. Although it might seem counterintuitive at first sight, a closer look at the model

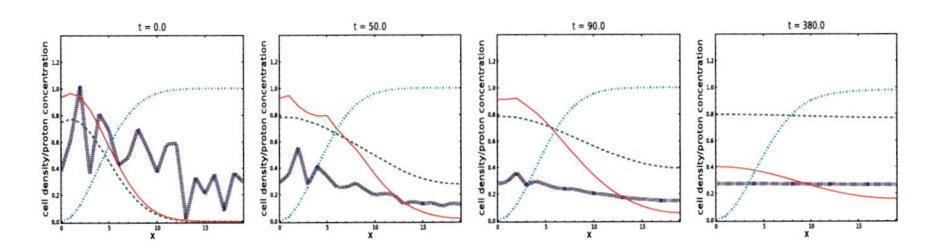

Fig. 3.2 Time snapshots for expected values of the SMAMCI model simulated in 1D over the space interval [0, 1.9]. H_i: blue curve of vertical bars, H_e: green, dashed curve, C: red, continuous curve, and N: cyan, dot-dashed curve. Reprinted from [31] with permission from Elsevier

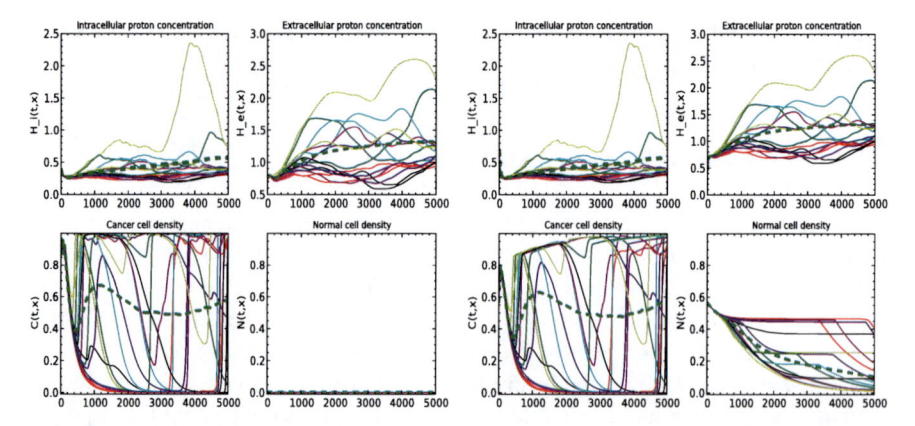

Fig. 3.3 Time evolution plots of the SMAMCI model for two different spatial points. Each of the two subfigures consists of four plots showing the evolution of H_i, H_e, C, and N. Thin, continuous curves: Sample paths of the solution. Thick, dashed curve: Mean sample path. The left subfigure corresponds to position $x = 0$, the right one to position $x = 0.5$. Reprinted from [31] with permission from Elsevier

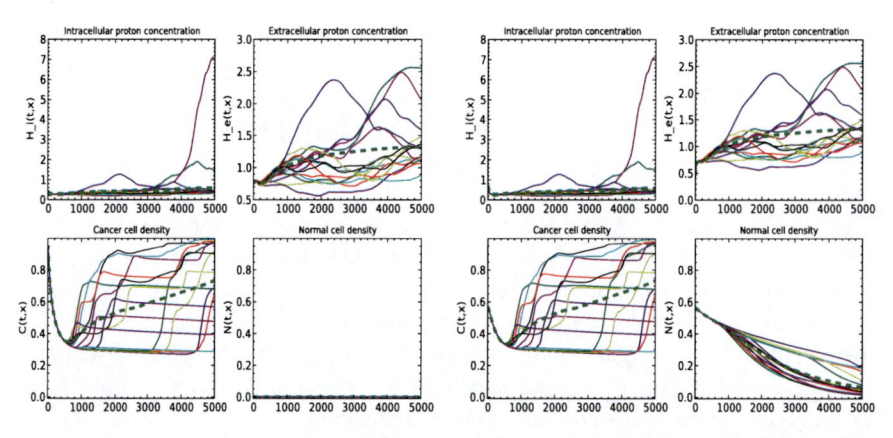

Fig. 3.4 Time evolution plots of the SMAMCI model at 2 different spatial points. The proliferation sensitivity parameter was taken to be relatively smaller. Thin, continuous curves: Sample paths of the solution. Thick, dashed curve: Mean sample path. The left subfigure corresponds to position $x = 0$ and the right one to position $x = 0.5$. Reprinted from [31] with permission from Elsevier

gives us an explanation: a smaller sensitivity parameter in Λ_1 reduces the growth rate, but also scales down the decay rate; as a result, a relatively smaller number of sample paths exhibit decay. Consequently, the expected cancer density is relatively high. (3) Because of the switching property of $\Lambda_1(H_i)$ the apparent extinction time of normal cells is significantly delayed.

3.3.3.2 Simulations for the SAIG Model

1D Simulations for the SAIG Model

Time sequences of different solution sample paths are shown in Figs. 3.5a and 3.6, from which the following can be inferred:

1. The dynamics of H_i is more prominent at the tumor periphery, i.e. at the interface of the tumor and stroma. Furthermore, the value of H_i at this region is relatively larger than the H_i values at the bulk region of the tumor. Phenomenologically, this relates to high cellular activity needed to meet the energy requirements of cytoskeleton remodeling (which facilitates cellular re-orientation), thus resulting in increased H_i production.
2. The experimentally observed behavior of reverse pH gradient can be seen in all sample solutions, which have H_e greater than H_i primarily over the bulk of tumor. However, noise fluctuations may temporarily cause the H_i concentration to exceed that of H_e.
3. The peak in the H_e concentration curve is mainly at the interface between cancer and normal cells. This is in good agreement with the phenomenon of acid accumulation at the tumor-stroma interface, due to the high glycolytic rate of peripheral tumor cells. This in turn has the following influence on the population level dynamics:

    3a. The concentrations of H_e and H_i control the behavioral mode of cancer cells, i.e. if the cancer cells are in the go, grow or recede mode. When H_i and H_e take values in the support of b, then the cancer cells are in the go-mode, consequently they exhibit pH-taxis. If H_i and H_e take values in the support of Λ_1, then the cancer cells are in the grow mode, therefore they proliferate. Finally, if the H_i and H_e take values in the support of Λ_2, the cancer cells are in the recede mode, so the tumor cell density decays.
    3b. Since the decay of normal cells is tightly coupled to the cancer density C and the H_e concentration, the interplay between the decay function Λ_3, the remodeling function Λ_4, and the go-grow-recede mode of cancer cells has a strong influence on the opening and closing of gaps.

To get a more detailed insight into the respective dynamics we look at different sample paths of the 1D simulations of the SAIG model:

1. In Fig. 3.5a, the time snapshots of the 90th sample solution, we observe an alternating sequence (with respect to time) of opening and closing of gaps. At $t = 130$ a gap is about to be formed (as seen by a ∨-shaped profile formed by the intersection of cancer and normal cell density curves). Due to the accumulated acid, the gap widens at around time $t = 220$. However, due to the invasion of cancer cells the gap begins to shrink and closes at about $t = 455$. The proliferating edge of the cancer in turn results in acid being accumulated at the interface, consequently leading to the reappearance of the gap (at time

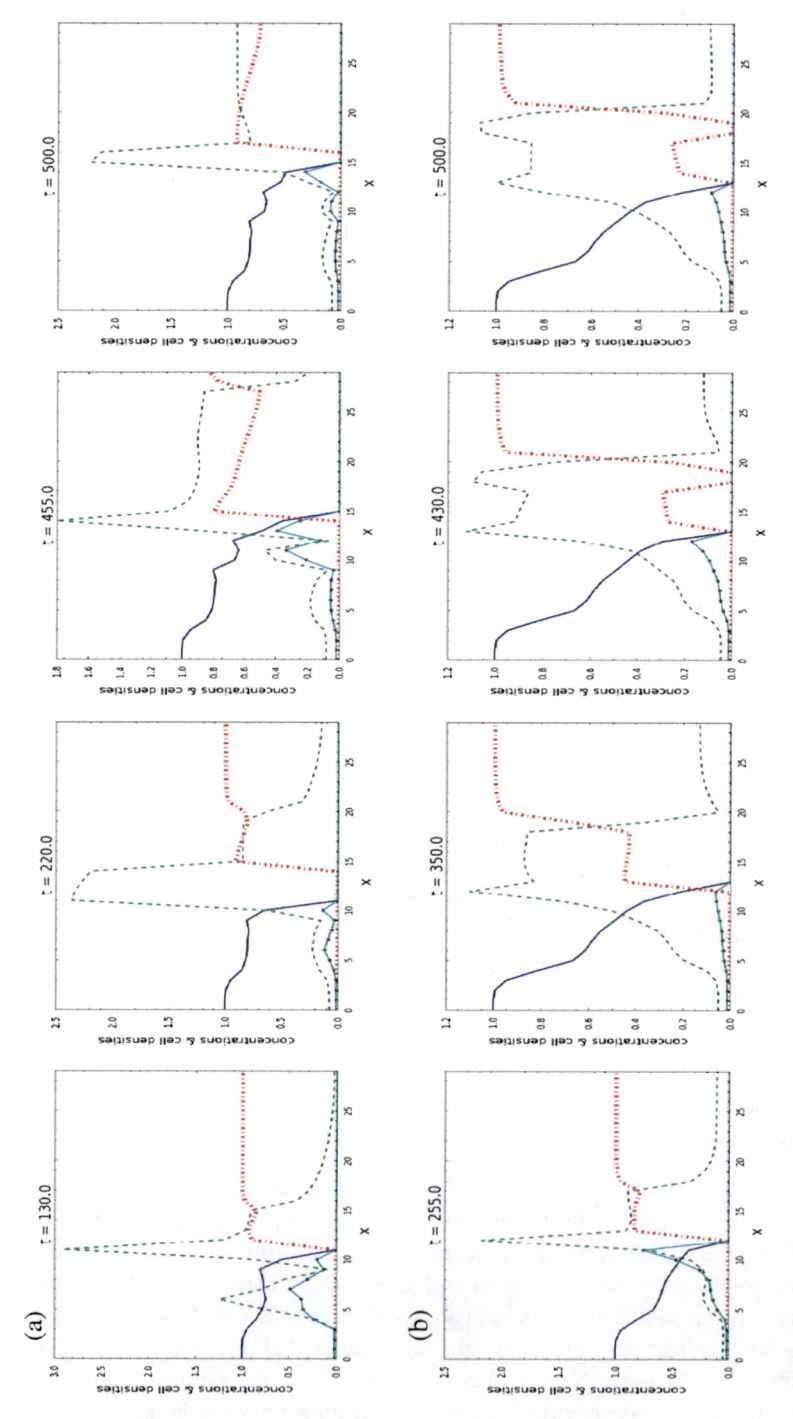

Fig. 3.5 Time snapshots of the sample solutions 90 and 63, in the case of a 1D domain for the SAIG model. Blue solid curve: cancer cell density C; green dashed curve: extracellular proton concentration H_e; red curve with vertical bars: normal cell density N, and cyan solid line with asterisks: intracellular proton concentration H_i. Reprinted from [32], © IOP Publishing & London Mathematical Society. Reproduced with permission. All rights reserved. (**a**) Sample solution 90. (**b**) Sample solution 63

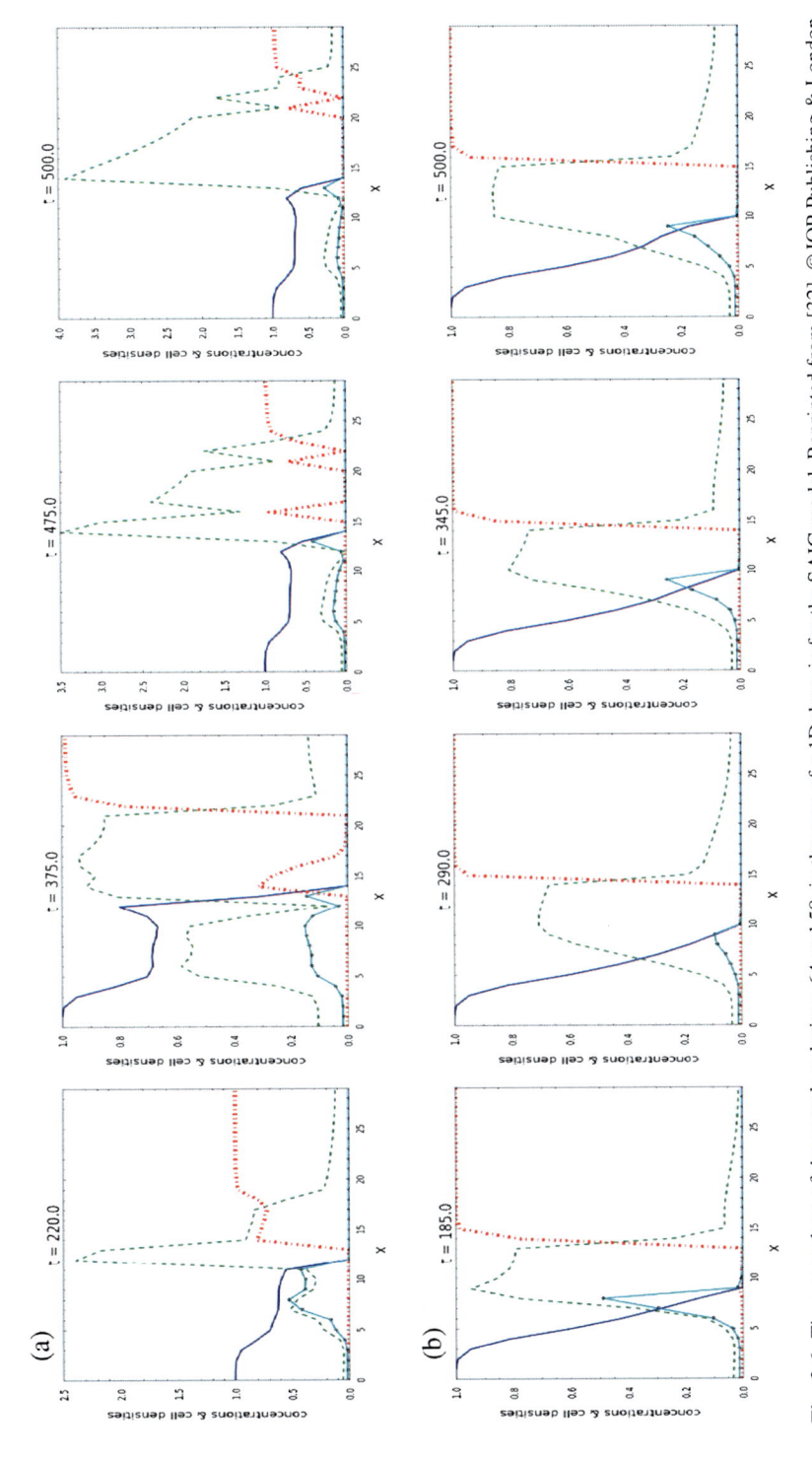

Fig. 3.6 Time snapshots of the sample solutions 64 and 58, in the case of a 1D domain for the SAIG model. Reprinted from [32], © IOP Publishing & London Mathematical Society. Reproduced with permission. (**a**) Sample solution 64. (**b**) Sample solution 58

$t = 500$). Such alternating sequence of gap/no-gap is observed for most of the sample paths.
2. In Fig. 3.5b showing some time snapshots of the 63rd sample solution we observe the appearance of a secondary gap (at time $t = 430$), which results in an isolated patch (island) of normal cells. The interplay between competing growth and decay (Λ_4 and Λ_3) terms in normal cell dynamics not only facilitates the growth of such isolated patches, but is also one of the probable reasons for the occurrence of infiltrative patterns.
3. In Fig. 3.6b showing time snapshots of the 58th sample solution we observe the formation of a large gap which is mainly due to the cancer cells being in the recede-mode. This is again due to the fact that the H_i and H_e take values in the support of the recede function. The gap widens as time progresses until the end of the simulation.
4. Lastly, in Fig. 3.7a showing a time sequence of plots for the expected solution (i.e., the numerical mean of the sample solutions) no gaps are begin formed. This is solely due to the fact that gaps (if they occur) appear at different positions, are of varying widths, and remain open for varying time spans. Therefore, on average no gaps are seen. That being said, by increasing and decreasing the normal cell decay rate and the normal cell remodelling rate, respectively, the formation of wider gaps can be facilitated, thereby increasing the probability of gap occurrence even in the averaged solution (Fig. 3.7b).

2D Simulations for the SAIG Model

To highlight the novelty and richness of the SAIG model we next present some simulation results in the 2D case. The plots show solid curves indicating the level sets of cancer cell density C overlapped on the colored contour regions of normal cell density N. The white regions appearing in the center are the gaps (regions of nearly zero cell density ($< 10^{-7}$)). The gaps and most of the patterns (islands, cavities) mentioned above in the 1D case also occur the 2D simulations. Even more interesting patterns can be observed, like the appearance of buds and spikes (finger-like projections, which are the source for the name 'cancer', due to the similarity in appearance with the shape of crab pliers). For a detailed description of the results we refer the reader to the original paper [32]. Here, for the sake of conciseness, we only depict the plots illustrating the emergence of INFb and INFc patterns. Infiltrative (INF) patterns are a way to classify the local invasiveness or the local malignancy of the cancer cells, as introduced by the Japanese gastric cancer association group [35]. The INFa pattern (see Fig. 3.8) is identified with the appearance of gaps between the cancer and stromal cells, indicating an early stage of an invading cancer. On the other hand, INFb and INFc mark the onset of malignancy where cancer cells are profusely intermixed with the normal cells. This is illustrated in Fig. 3.9, where the invading front of the tumor is overlapping with the stromal region. Another interesting observation is

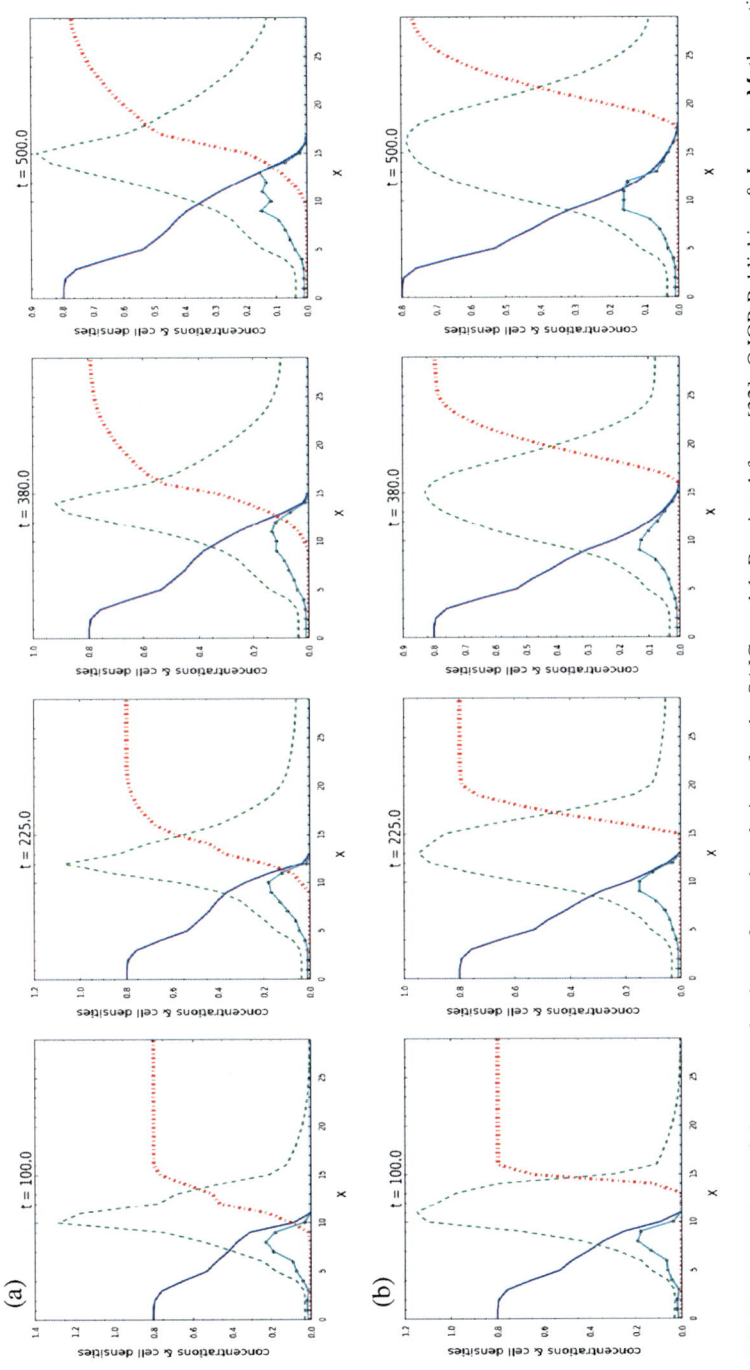

3 Mathematical Models for Acid-Mediated Tumor Invasion: From... 63

Fig. 3.7 Time snapshots of the expected values of sample solutions for the SAIG model. Reprinted from [32], © IOP Publishing & London Mathematical Society. Reproduced with permission. All rights reserved. (**a**) Numerical expectation of sample solutions. (**b**) Numerical expectation of sample solutions (SAIG model) for a relatively higher decay parameter associated to H_e

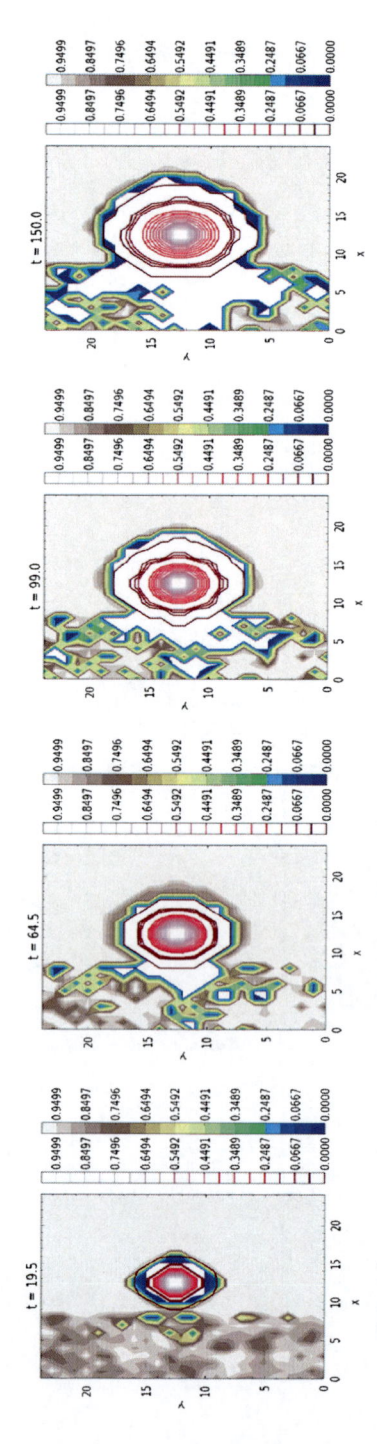

Fig. 3.8 Simulation plots for the SAIG model depicting the INFa patterns. The solid curves indicate the level sets of cancer cell density C, while filled regions indicate level sets of normal cell density N. The values corresponding to these level sets are indicated by the color bars adjacent on the right side to the 2D plots. In order to observe the effects of spatial heterogeneity, we added some random perturbation to the initial value of the normal cell density only on the left side of the xy-plane. Thus, these perturbations are seen as patches on the left half of the 2D plots (see also Fig. 3.9). Reprinted from [32], © IOP Publishing & London Mathematical Society. Reproduced with permission. All rights reserved

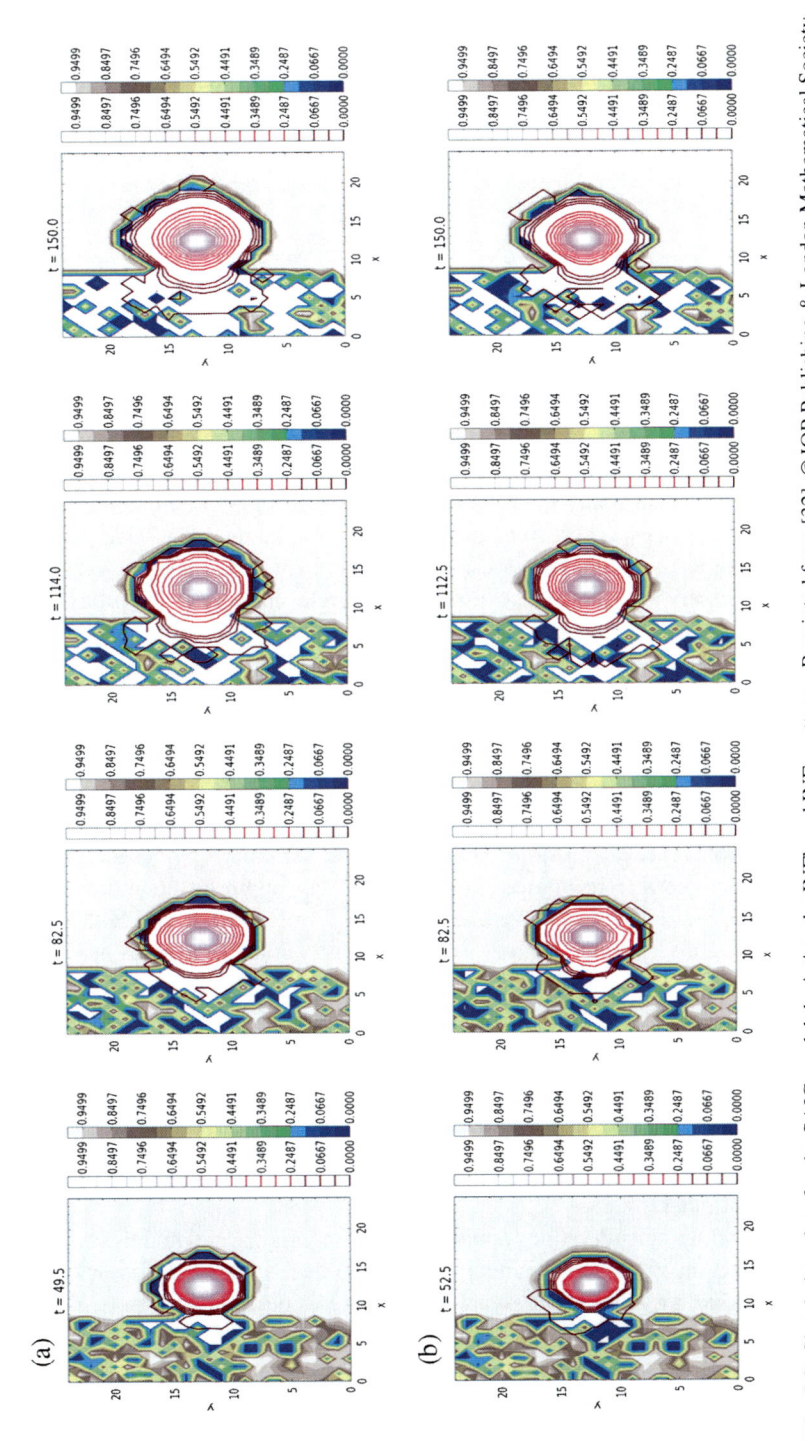

Fig. 3.9 Simulation plots for the SAIG model depicting the INFb and INFc patterns. Reprinted from [32], © IOP Publishing & London Mathematical Society. Reproduced with permission. All rights reserved. (**a**) Sample solution 2. (**b**) Sample solution 3

that, due to the two sided nature of the stroma region (left heterogeneous part versus the right homogeneous part), the spread of cancer is highly skewed. Due to the sparsity of the tissue on left side, the cancer cells have a bias to move towards this region mainly mediated by the spread of acid in the empty sites of the tissue. This mainly highlights the fact that spatial heterogeneity and spatial randomness have a strong influence on the emergence of INF patters and the invasiveness of cancer.

3.4 Discussion

We reviewed in this chapter some of the main directions followed so far in the mathematical description of acid-mediated tumor invasion, thereby focusing on the continuum formulations. Although the settings have persistently improved and are able to account for many important features of this highly complex biological process, a lot of work has still to be done, not only on the modeling side, but also with respect to analysis and numerics. The biology is not fully understood either, and it is necessary to identify the aspects most relevant for tumor progression and invasion. While there is vast knowledge available for macroscopic PDE systems coupling equations of the same type, connecting PDEs with ODEs still leads to serious difficulties from the analytical point of view, in particular when nonlinear, possibly degenerate diffusion and different types of taxis are accounted for, even in a pure macroscopic context. Including multiple scales contributes to enhancing the predictive power of the models, however renders their mathematical handling more involved and leaves many related issues (e.g., well-posedness in less regular function spaces, long time behavior of the solution, uniqueness, correct and efficient numerical techniques) open. This also applies (the more so) to stochastic continuum models, where even less informations are available about well-posedness of reaction-diffusion-taxis equations coupled to RODEs and SDEs acting on different time and space scales and where positivity-preserving schemes for handling such systems have yet to be developed. Actually, only few of the mathematical challenges are specific to the problem of modeling acid-mediated tumor invasion; rather, the biological problem gives rise to models which are of theoretical interest on their own. The settings can be extended to capture even further clinically relevant aspects, like lymph- and blood angiogenesis, haptotaxis combined with pH-taxis, various treatment approaches combining (neo)adjuvant chemo and radiotherapy, etc.

While the above considerations apply to mono- and multiscale models involving subcellular and macroscopic dynamics, one has to deal with similar issues when it comes to coupling lower scale dynamics, aiming to describe in a more careful way biological processes on the subcellular scale (ODEs/SDEs/RODEs) interacting with cell density functions on the mesoscopic scale, i.e., depending not only on time and position, but also on the velocity of the cells and their internal state (here concentration of intracellular protons) and with some macroscopic quantity (concentration of extracellular protons). The mathematical analysis of such systems coupling ODEs, Boltzmann-like kinetic transport equations, and reaction-diffusion equations is involved and not seldom a challenge, particularly if less generous assumptions are made about the data. We refer to [41] for the global well-posedness of such system, in a different, but very closely related context. The model studied there features both chemotaxis and haptotaxis, thus being more general than a (multiscale) kinetic model for pH-taxis and cell-tissue interactions. Numerical simulations of such multiscale models involving one or several integro-differential transport equations are still out of reach; instead, it is desirable to deduce effective macroscopic equations from such models by adequate scalings and moment closure. This not only allows to efficiently simulate the behavior of the tumor, thereby still accounting for manifold effects, but also provides insights into the precise form of the terms and coefficients in the corresponding equations on the population scale, thus reducing the arbitrariness of assumptions on the macroscopic level. However, whereas single equations set on the mesoscopic level can often be scaled (although rather formally), when such equations are coupled in a nonlinear way it is not clear how to perform the macroscopic scaling, nor is the moment closure analytically established. As mentioned before, the alternative approach of starting from SDEs for the cell position on the individual cell level and directly deducing macroscopic equations (without passing through the mesolevel) has led in a related context to rigorous results, e.g., in [47, 48]. The more delicate case with SDEs describing position, velocity, and internal states of the cells on the individual cell scale and the adequate way of deducing the macroscopic cell dynamics e.g., by some mean-field limit is—if available at all—still formal, the limits being not rigorously established. We are not aware of any such results related to tumor invasion and hypoxia.

In conclusion, there seems to be a need of new methods or at least of new ideas in order to handle problems of such complexity. Including stochasticity in an explicit manner in the continuum models—although important in order to enhance the description—increases the degree of difficulty and there are hardly references dealing with such problems, not only in the context of (acid-mediated) cancer invasion.

Appendix

Definition 3.1 (Bessel Potential Space on $\mathbb{R}^n$ (See [32])) The space $H_p^s(\mathbb{R}^n)$ with $s \in \mathbb{R}$ and $1 < p < \infty$ is the function space defined by

$$H_p^s(\mathbb{R}^n) := \left\{ f \in \mathscr{S}'(\mathbb{R}^n) : \mathfrak{F}^{-1}\left[(1 + |\xi|^2)^{\frac{s}{2}} \mathfrak{F} f\right] \in L^p(\mathbb{R}^n) \right\}$$

$$\|f\|_{H_p^s(\mathbb{R}^n)} := \|\mathfrak{F}^{-1}\left[(1 + |\xi|^2)^{\frac{s}{2}} \mathfrak{F} f\right]\|_{L^p(\mathbb{R}^n)}.$$

where $\mathscr{S}'(\mathbb{R}^n)$ denotes the dual of the Schwartz space:

$$\mathscr{S}(\mathbb{R}^n) := \{u \in C^\infty(\mathbb{R}^n) \ : \ \forall \, \alpha, \, \beta \in \mathbb{N}^n \ \sup_{\mathbf{x} \in \mathbb{R}^n} |\mathbf{x}^\alpha D^\beta u(\mathbf{x})| < \infty\}$$

and $\mathfrak{F}$ is the usual Fourier transform operator.

Definition 3.2 (Bessel Potential Space on $\mathfrak{D}$ (See [32])) Let $\mathfrak{D}$ be a bounded Lipschitz domain in $\mathbb{R}^n$. $u \in H_p^s(\mathfrak{D})$ is an equivalence class of functions $U \in H_p^s(\mathbb{R}^n)$ such that $U|_\mathfrak{D} = u$. That is a function u is said to be in $H_p^s(\mathfrak{D})$ if and only if there exits a function $U \in H_p^s(\mathbb{R}^n)$ such that $U|_\mathfrak{D} = u$.

Moreover, $H_p^s(\mathfrak{D})$ endowed with the norm

$$\|u\|_{H_p^s(\mathfrak{D})} := \inf_{\substack{U|_\mathfrak{D}=u, \\ U \in H_p^s(\mathbb{R}^n)}} \|U\|_{H_p^s(\mathbb{R}^n)}$$

is a Banach space.

References

1. Abakarova O (1995) The metastatic potential of tumors depends on the pH of host tissues. Bull Exp Biol Med 120:1227–1229
2. Al-Husari M, Webb S (2013) Acid-mediated tumour cell invasion: a discrete modelling approach using the extended potts model. Clin Exp Metastasis 30:793–806
3. Al-Husari M, Webb S (2013) Regulation of tumour intracellular pH: a mathematical model examining the interplay between H^+ and lactate. J Theor Biol 322:58–71
4. Bartel P, Ludwig F, Schwab A, Stock C (2012) pH-taxis: directional tumor cell migration along pH-gradients. Acta Physiol 204:113
5. Beckner M, Stracke M, Liotta L, Schiffmann E (1990) Glycolysis as primary energy source in tumor cell chemotaxis. J Natl Cancer Inst 82(23):1836–1840
6. Boyer M, Tannock I (1992) Regulation of intracellular pH in tumor cell lines: influence of microenvironmental conditions. Cancer Res 52(16):4441–4447. http://cancerres.aacrjournals.org/content/52/16/4441.full.pdf+html
7. Burbridge M, West D, Atassi G, Tucker G (1999) The effect of extracellular pH on angiogenesis in vitro. Angiogenesis 3:281–288
8. Byrne HM (2012) Mathematical biomedicine and modeling avascular tumor growth. In: Mathematics and life sciences. De Gruyter (submitted). http://eprints.maths.ox.ac.uk/1647/1/finalOR96.pdf
9. Casey J, Grinstein S, Orlowski J (2010) Sensors and regulators of intracellular pH. Nat Rev Mol Cell Biol 11(1):50–61. http://dx.doi.org/10.1038/nrm2820
10. Chaplain M, Lolas G (2005) Mathematical modelling of cancer cell invasion of tissue: the role of the urokinase plasminogen activation system. Math Models Methods Appl Sci 15:1685–1734
11. Charman R (1996) Electrical properties of cells and tissues. In: Kitchen S, Bazi S (eds) Clayton's electrotherapy, 10th edn. WB Saunders, London
12. Chavanis PH (2010) A stochastic Keller–Segel model of chemotaxis. SI Chaos Complexity Transp Theory Appl 15:60–70

13. Corcoran A, Del Maestro R (2003) Testing the "go or grow" hypothesis in human medulloblastoma cell lines in two and three dimensions. Neurosurgery 53:174–185
14. De Milito A, Fais S (2005) Tumor acidity, chemoresistance and proton pump inhibitors. Future Oncol 1:779–786
15. Dhup S, Dadhich R, Porporato P, Sonveaux P (2012) Multiple biological activities of lactic acid in cancer: influences on tumor growth, angiogenesis and metastasis. Curr Pharm Des 18:1319–1330
16. Engwer C, Hillen T, Knappitsch M, Surulescu C (2015) Glioma follow white matter tracts: a multiscale DTI-based model. J Math Biol 71:551–582
17. Evans L (1997) Linear evolution equations, chap. 7. In: Partial differential equations, vol 49. American Mathematical Society, Providence, RI
18. Fasano A, Herrero M, Rocha R (2009) Slow and fast invasion waves in a model of acid-mediated tumour growth. Math Biosci 220:45–56
19. Freeman M, Sierra E (1984) An acidic extracellular environment reduces the fixation of radiation damage. Radiat Resist 97:154–161
20. Ganapathy-Kanniappan S, Geschwind JF (2013) Tumor glycolysis as a target for cancer therapy: progress and prospects. Mol Cancer 12:152–163
21. Garay T, Juhász E, Molnàr E, Eisenbauer M, Czirók A, Dekan B, László V, Hoda M, Döme B, Tímár J, Klepetko W, Berger W, Hegedűs B (2013) Cell migration or cytokinesis and proliferation? Revisiting the "go or grow" hypothesis in cancer cells in vitro. Exp Cell Res 319(20):3094–3103
22. Gatenby R, Gawlinski E (1996) A reaction-diffusion model of cancer invasion. Cancer Res 56(24):5745–5753. http://cancerres.aacrjournals.org/content/56/24/5745.full.pdf+html
23. Gatenby R, Gawlinski E (2003) The glycolytic phenotype in carcinogenesis and tumor invasion: insights through mathematical models. Cancer Res 63(14):3847–3854. http://cancerres.aacrjournals.org/content/63/14/3847.full.pdf+html
24. Gatenby RA, Gillies R (2004) Why do cancers have high aerobic glycolysis? Nat Rev Cancer 4(11):891–899. http://dx.doi.org/10.1038/nrc1478
25. Gerlee P, Anderson A (2008) A hybrid cellular automaton model of clonal evolution in cancer: the emergence of the glycolytic phenotype. J Theor Biol 250:705–722
26. Giese A, Loo M, Tran N, Haskett D, Coons S, Berens M (1996) Dichotomy of astrocytoma migration and proliferation. Int J Cancer 67:275–282
27. Gillies R, Martinez-Zaguilan R, Peterson E, Perona R (1992) Role of intracellular pH in mammalian cell proliferation. Cell Physiol Biochem 2:159–179
28. Granchi C, Fancelli D, Minutolo F (2014) An update on therapeutic opportunities offered by cancer glycolytic metabolism. Bioorg Med Chem Lett 24:4915–4925
29. Hanahan D, Weinberg R (2011) Hallmarks of cancer: the next generation. Cell 144(5):646–674. http://dx.doi.org/10.1016/j.cell.2011.02.013
30. Harris A (2002) Hypoxia—a key regulatory factor in tumour growth. Nat Rev Cancer 2:38–47
31. Hiremath S, Surulescu C (2015) A stochastic multiscale model for acid mediated cancer invasion. Nonlinear Anal Real World Appl 22(0):176–205
32. Hiremath S, Surulescu C (2016) A stochastic model featuring acid-induced gaps during tumor progression. Nonlinearity 29(3):815
33. Holder A, Rodrigo M, Herrero M (2014) A model for acid-mediated tumour growth with nonlinear acid production term. Appl Math Comput 227:176–198
34. Jakobsson E, Chiu S (1987) Stochastic theory of ion movement in channels with single-ion occupancy. Application to sodium permeation of gramicidin channel. Biophys J 52:33–45
35. Japanese Gastric Cancer Association (2011) Japanese classification of gastric carcinoma: 3rd english edition. Gastric Cancer 14(2):101–112
36. Kato Y, Ozawa S, Miyamoto C, Maehata Y, Suzuki A, Maeda T, Baba Y (2013) Acidic extracellular microenvironment and cancer. Cancer Cell Int 13(24):89
37. Kloeden P, Sonner S, Surulescu C (2016) A nonlocal sample dependence SDE-PDE system modeling proton dynamics in a tumor. Dyn Syst Ser B 21(7):2233–2254. http://dx.doi.org/10.3934/dcdsb.2016045

38. Kunkel M, Reichert T, Benz Pea (2003) Overexpression of glut-1 and increased glucose metabolism in tumors are associated with a poor prognosis in patients with oral squamous cell carcinoma. Cancer 97(4):1015–1024. http://dx.doi.org/10.1002/cncr.11159
39. Lee A, Tannock I (1998) Heterogeneity of intracellular pH and of mechanisms that regulate intracellular pH in populations of cultured cells. Cancer Res 58(9):1901–1908
40. Lee HS, Park H, Lyons J, Griffin R, Auger E, Song C (1997) Radiation-induced apoptosis in different pH environments in vitro. Int J Radiat Oncol Biol Phys 38:1079–1087
41. Lorenz T, Surulescu C (2014) On a class of multiscale cancer cell migration models: well-posedness in less regular function spaces. Math Models Methods Appl Sci 24:2383–2436
42. Martinez-Zaguilan R, Seftor E, Seftor R, Chu Y, Gillies R, Hendrix M (1996) Acidic pH enhances the invasive behavior of human melanoma cells. Clin Exp Metastasis 14:176–186
43. McGillen J, Martin N, Robey I, Gaffney E, Maini P (2012) Application of mathematical analysis to tumor acidity modeling. RIMS Kokyuroku Bessatsu B31:31–59
44. McGillen J, Gaffney E, Martin N, Maini P (2014) A general reaction–diffusion model of acidity in cancer invasion. J Math Biol 68(5):1199–1224
45. Meral G, Stinner C, Surulescu C (2015) A multiscale model for acid-mediated tumor invasion: therapy approaches. J Coupled Syst Multiscale Dyn 3:135–142
46. Mochiki E, Kuwano H, Katoh H, Asao T, Oriuchi N, Endo K (2004) Evaluation of 18f-2-deoxy-2-fluoro-d-glucose positron emission tomography for gastric cancer. World J Surg 28:247–253
47. Oelschlaeger K (1989) On the derivation of reaction-diffusion equations as limit dynamics of systems of moderately interacting stochastic processes. Probab Theory Relat Fields 82:565–586
48. Othmer H, Stevens A (1997) Aggregation, blowup and collapse: the abc's of taxis in reinforced random walks. SIAM J Appl Math 57:1044–1081
49. Paradise R, Whitfield M, Lauffenburger D, Van Vliet K (2013) Directional cell migration in an extracellular pH gradient: a model study with an engineered cell line and primary microvascular endothelial cells. Exp Cell Res 319:487–497
50. Patel A, Gawlinski E, Lemieux S, Gatenby R (2001) Cellular automaton model of early tumor growth and invasion: the effects of native tissue vascularity and increased anaerobic tumor metabolism. J Theor Biol 213:315–331
51. Pavlin M, Pavselj N, Miklavcic D (2002) Dependence of induced transmembrane potential on cell density arrangement, and cell position inside a cell system. IEEE Trans Biomed Eng 49:605–612
52. Pazy A (1983) Semigroups of linear operators and applications to partial differential equations. Applied Mathematical Sciences. Springer, New York
53. Rofstad EK, Mathiesen B, Kindem K, Galappathi K (2006) Acidic extracellular pH promotes experimental metastasis of human melanoma cells in athymic nude mice. Cancer Res 66(13):6699–6707
54. Sachs R, Brenner D (1998) The mechanistic basis of the linear-quadratic model. Med Phys 25:2071–2073
55. Shrode L, Tapper H, Grinstein S (1997) Role of intracellular pH in proliferation, transformation, and apoptosis. J Bioenergetics Biomembranes 29(4):393–399. http://dx.doi.org/10.1023/A:1022407116339
56. Smallbone K, Gavaghan D, Gatenby R, Maini P (2005) The role of acidity in solid tumor growth and invasion. J Theor Biol 235:476–484
57. Smallbone K, Gatenby R, Gillies RJ, Maini P, Gavaghan D (2007) Metabolic changes during carcinogenesis: potential impact on invasiveness. J Theor Biol 244:703–713
58. Smallbone K, Gatenby R, Maini P (2008) Mathematical modelling of tumour acidity. J Theor Biol 255:106–112
59. Song C, Griffin R, Park H (2006) Influence of tumor pH on therapeutic response. In: Teicher B (ed) Cancer drug resistance. Humana Press, Totowa, NJ
60. Stinner C, Surulescu C, Meral G (2015) A multiscale model for pH-tactic invasion with time-varying carrying capacities. IMA J Appl Math 80:1300–1321

61. Stinner C, Surulescu C, Uatay A (2016) Global existence for a go-or-grow multiscale model for tumor invasion with therapy. Math Models Methods Appl Sci 26:2163. https://doi.org/10.1142/S021820251640011X
62. Stock C, Schwab A (2009) Protons make tumor cells move like clockwork. Eur J Physiol 458:981–992
63. Stokes C, Lauffenburger D, Williams S (1991) Migration of individual microvessel endothelial cells: stochastic model and parameter measurement. J Cell Sci 99:419–430
64. Stubbs M, McSheehy P, Griffiths J, Bashford L (2000) Causes and consequences of tumour acidity and implications for treatment. Mol Med Today 6(1):15–19. http://dx.doi.org/10.1016/S1357-4310(99)01615-9
65. Swanson K, Rockne R, Claridge J, Chaplain Jr M, Alvord E, Anderson A (2011) Quantifying the role of angiogenesis in malignant progression of gliomas: in silico modeling integrates imaging and histology. Cancer Res 71:7366–7375
66. Tannock I, Rotin D (1989) Acid pH in tumors and its potential for therapeutic exploitation. Cancer Res 49:4373–4384
67. Tao Y, Wang M (2008) Global solution for a chemotactic–haptotactic model of cancer invasion. Nonlinearity 21:2221–2238
68. Tao Y, Winkler M (2014) Dominance of chemotaxis in a chemotaxis–haptotaxis model. Nonlinearity 27:1225–1239
69. Van der Heiden M, Cantley L, Thompson C (2009) Understanding the Warburg effect: the metabolic requirements of cell proliferation. Science 324(5930):1029–1033. http://dx.doi.org/10.1126/science.1160809. http://www.sciencemag.org/content/324/5930/1029.full.pdf
70. Webb S, Sherratt J, Fish R (1999) Alterations in proteolytic activity at low pH and its association with invasion: a theoretical model. Clin Exp Metastasis 17(5):397–407
71. Webb S, Sherratt J, Fish R (1999) Mathematical modelling of tumor acidity: regulation of intracellular pH. J Theor Biol 196(2):237–250. http://dx.doi.org/10.1006/jtbi.1998.0836
72. Webb B, Chimenti M, Jacobson M, Barber D (2011) Dysregulated pH: a perfect storm for cancer progression. Nat Rev Cancer 11(9):671–7
73. Zheng P, Severijnen L, van der Weiden M, Willemsen R, Kros J (2009) Cell proliferation and migration are mutually exclusive cellular phenomena in vivo: implications for cancer therapeutic strategies. Cell Cycle 8:950–951
74. Zhu A, Lee D, Shim H (2011) Metabolic pet imaging in cancer detection and therapy response. Semin Oncol 38:55–69

Chapter 4
Numerical Simulation of a Contractivity Based Multiscale Cancer Invasion Model

Niklas Kolbe, Mária Lukáčová-Medvid'ová, Nikolaos Sfakianakis, and Bettina Wiebe

4.1 Introduction

The primary objectives in cancer research are to understand the causes of cancer in order to develop strategies for its diagnosis and treatment. The overall effort involves the medical science, biology, chemistry, physics, computer science, and mathematics. The contribution of mathematics, in particular, spans from the modelling of the relevant biological processes, to the analysis of the developed models, and their numerical simulations. The range of applications of the mathematical models covers a wide range of processes from intracellular bio-chemical reactions to cancer growth, its metastasis and treatment, e.g. [1, 3, 4, 10, 11, 13, 14, 16, 22, 25–28, 30, 31, 35, 37, 39].

In this work we focus on the first step of cancer metastasis—and one of the "hallmarks of cancer"—the invasion of the *extracellular matrix* (ECM). Our study involves the existence of a secondary group of cancer cells within the main body of the tumour that exhibits *stem-cell-like* properties. This secondary group of cancer cells seem to stem from the "original" cancer cells via a cellular differentiation program that can be found also in normal tissue, the *Epithelial-Mesenchymal Transition* (EMT). Both the EMT and its reverse process, *Mesenchymal-Epithelial Transition*

N. Kolbe · M. Lukáčová-Medvid'ová · B. Wiebe
Institute of Mathematics, Johannes Gutenberg-University, Staudingerweg 9, 55128 Mainz, Germany
e-mail: kolbe@uni-mainz.de; lukacova@uni-mainz.de; b.wiebe@uni-mainz.de

N. Sfakianakis (✉)
Institute of Applied Mathematics, Heidelberg University, Im Neuenheimer Feld 205, 69120 Heidelberg, Germany
e-mail: sfakiana@math.uni-heidelberg.de

© Springer International Publishing AG, part of Springer Nature 2017
A. Gerisch et al. (eds.), *Multiscale Models in Mechano and Tumor Biology*,
Lecture Notes in Computational Science and Engineering 122,
https://doi.org/10.1007/978-3-319-73371-5_4

(MET) participate in several developmental processes including embryogenesis, wound healing, and fibrosis, [12, 17, 23, 34, 38].

The two types of cancer cells possess different cell *proliferation* rates and motility properties, and present different levels of cellular *potency*. The secondary group, in particular, exhibits lower (if any) proliferation rates, stem cell-like properties such as self-renewal and cellular differentiation. These cells are more resilient to cancer therapies and they are able to metastasise. While the bulk consists mostly of the "original" cancer cells, the secondary family constitutes the smaller part of the tumour, [14, 32].

The motility mechanism of the cancer cells responds to alterations and gradients in the chemical environment of the tumour (a process termed *chemotaxis*), and in the cellular adhesions sites located on the ECM (a process termed *haptotaxis*). From a mathematical perspective the study of several forms of -taxis has been an active research field in the last decades. The derived models are typically *Keller-Segel* (KS) type systems [18, 29], where the participating quantities are described macroscopically in the sense of densities. By including also interactions between the cancer cells and the extracellular environment, the resulting models take the form of *Advection-Reaction-Diffusion* (ARD) systems, see e.g. [1–3, 5, 9, 15, 33, 36].

The solutions of these models exhibit typically complex dynamical behaviour manifested in the form of merging/emerging concentration or in the form of complex wave phenomena [6, 9, 33]. Moreover, since these models are close to the classical KS systems, the possibility of a blow-up—if not analytically excluded—should be numerically investigated. Due to such dynamical behaviours, special and problem specific numerical treatments are needed, [13, 15, 20].

In the current paper our aim is to contribute in this direction by presenting our problem-suited numerical method for the solution of a particular ARD cancer invasion haptotaxis model that was proposed in [36]. This model features several numerically challenging properties: non constant advection and diffusion coefficients, non-local time delay, and stiff reaction terms, see (4.3).

The rest of the paper is structured as follows: in Sect. 4.2 we present and discuss briefly the cancer invasion model. In Sect. 4.3 we address the numerical method we employ, comment on its properties, and on the special treatment of its terms. In Sect. 4.4 we present our numerical findings and discuss their implication in terms of the model.

4.2 Mathematical Model

The model we investigate is a cancer invasion ARD system of the KS spirit that primarily features two families of cancer cells and includes *contractivity*: a measure of the strength of the cell-matrix *migration*, see [36]. In some more details, the following properties are assumed by the model:

- The "original" cancer cells (henceforth *proliferative*) proliferate and do not migrate or otherwise translocate. The stem-like cancer cells (henceforth *migratory*) migrate but do not proliferate.
- The motility mechanism of the migratory cancer cells responds to the (possibly) non-uniform distribution of adhesion sites located on the ECM. The induced haptotactic movement is modelled by a combination of advection and diffusion.
- There exist a *bidirectional* transition between the two families of cancer cells, modelling the parallel action of EMT and MET. Both are assumed to take place with constant rates.
- The ECM is a dynamic structure that is degraded by the cancer cells, and constantly remodelled. The remodelling is mostly due to living "agents" like the *fibroblast* or *cancer-associated fibroblast* cells. For simplicity though, it is assumed in this model that the ECM remodelling is self-induced, i.e. new ECM sprouts from existing ECM.
- The proliferation of the cancer cells and the remodelling of the ECM is limited by the locally available space. This effect is modelled by a volume filling term.
- As the cancer cells attach on the ECM, new *adhesion* sites/*integrins* are created. They degrade with a constant rate and reproduce with a preferable maximum local density.
- Changes in the adhesion sites are reflected to the morphology and the motility apparatus of the cells, the properties of which are modelled by the contractivity. These changes do not happen instantaneously, some (delay) time elapses before the intracellular signalling cascade is completed and the morphological changes take place.
- Both the diffusion and the advection of the migratory cancer cells are governed by non-uniform coefficients depending on the contractivity.
- The dynamics of the integrins and the contractivity take place on a *microscopic time scale* that is faster than the *macroscopic time scale* of the dynamics of the cancer cells.

Altogether the model reads:

$$\begin{cases} \partial_t c_1 = \mu_c c_1 (1 - (c_1 + c_2) - \eta_1 v) + \gamma c_2 - \lambda c_1 \\ \partial_t c_2 = \nabla \cdot \left(D_c \dfrac{\kappa}{1 + (c_1 + c_2)v} \nabla c_2 \right) - \nabla \cdot \left(D_h \dfrac{\kappa v}{1 + v} c_2 \nabla v \right) + \lambda c_1 - \gamma c_2 \\ \partial_t v = -\delta_v (c_1 + c_2) v + \mu_v v (1 - \eta_2 (c_1 + c_2) - v) \\ \partial_\vartheta y = k_1 (1 - y) v - k_{-1} y \\ \partial_\vartheta \kappa = -q\kappa + My(\vartheta - \tau) \end{cases} \qquad (4.1)$$

where c_1, c_2 denote the densities of the proliferating and the migrating cancer cells respectively. The densities of the ECM and of the ECM-bound integrins are denoted by v and y. The contractivity is denoted by κ, and the microscopic and macroscopic time scales by ϑ, and t respectively. We assume that the time scales are related in the following way:

$$\vartheta = \frac{t}{\chi}, \tag{4.2}$$

where the rescaling factor $0 < \chi < 1$ is a fixed constant. The *time delay* $\tau > 0$ is also assumed to be a constant.

The system (4.1) is endowed with initial and boundary conditions, see Sect. 4.4.1.

Rescaled System Using the time scale relation (4.2) we can rescale the model (4.1) and obtain the following system using only the macroscopic time variable t:

$$\begin{cases} \partial_t c_1 = \mu_c c_1 (1 - (c_1 + c_2) - \eta_1 v) + \gamma c_2 - \lambda c_1 \\ \partial_t c_2 = \nabla \cdot \left(D_c \frac{\kappa}{1 + (c_1 + c_2) v} \nabla c_2 \right) - \nabla \cdot \left(D_h \frac{\kappa v}{1 + v} c_2 \nabla v \right) + \lambda c_1 - \gamma c_2 \\ \partial_t v = -\delta_v (c_1 + c_2) v + \mu_v v (1 - \eta_2 (c_1 + c_2) - v) \\ \partial_t y = \frac{k_1}{\chi} (1 - y) v - \frac{k_{-1}}{\chi} y \\ \partial_t \kappa = -\frac{q}{\chi} \kappa + \frac{M}{\chi} y(t - \chi \tau) \end{cases} \tag{4.3}$$

Operator Form The system (4.3) can be written for convenience in a compact operator form as follows:

$$\mathbf{w}_t = D(\mathbf{w}) - A(\mathbf{w}) + R(\mathbf{w}), \tag{4.4}$$

where $\mathbf{w} : \Omega \times \mathbb{R}_+ \to \mathbb{R}^5$, with $\mathbf{w} = (c_1, c_2, v, y, \kappa)^T$, and D, A, R represent the diffusion, advection, and reaction operators, respectively:

$$D(\mathbf{w}) = \left(0, \nabla \cdot \left(D_c \frac{\kappa}{1 + (c_1 + c_2) v} \nabla c_2 \right), 0, 0, 0 \right)^T,$$

$$A(\mathbf{w}) = \left(0, \nabla \cdot \left(D_H \frac{\kappa v}{1 + v} c_2 \nabla v \right), 0, 0, 0 \right)^T,$$

$$R(\mathbf{w}) = \begin{pmatrix} \mu_c c_1 (1 - (c_1 + c_2) - \eta_1 v) + \gamma c_2 - \lambda c_1 \\ \lambda c_1 - \gamma c_2 \\ -\delta_v (c_1 + c_2) v + \mu_v v (1 - \eta_2 (c_1 + c_2) - v) \\ \frac{k_1}{\chi} (1 - y) v - \frac{k_{-1}}{\chi} y \\ -\frac{q}{\chi} \kappa + \frac{M}{\chi} y(\vartheta - \tau) \end{pmatrix}.$$

Additionally we set

$$R_{\text{impl}}(\mathbf{w}) = \left(0,\ 0,\ 0,\ \frac{k_1}{\chi}(1-y)v - \frac{k_{-1}}{\chi}y,\ 0\right)^T, \quad (4.5)$$

and

$$R_{\text{expl}}(\mathbf{w}) = R(\mathbf{w}) - R_{\text{impl}}(\mathbf{w}). \quad (4.6)$$

Parameters For the main experiments we consider the following set of parameters that has been adopted by Stinner et al. [36]:

$$\begin{cases} \mu_c = 1, & \eta_1 = 0.05, & \gamma = 0.055, & \lambda = 0.152, \\ D_c = 0.01, & D_h = 10, \\ \delta_v = 5, & \mu_v = 0.3, & \eta_2 = 0.9, \\ k_1 = 2, & k_{-1} = 0.06, \\ q = 3, & M = 2, \\ \chi = 0.01, & \tau = 20. \end{cases} \quad (4.7)$$

These parameters are adjusted in each particular experiment under investigation, see also Sect. 4.4.1.

4.3 Numerical Method

An efficient and accurate numerical discretisation of cancer invasion models requires elaborate methods. In this work we follow a second order upwind approach for haptotaxis that prevents spurious oscillations and negative numerical results. Moreover, we apply an Implicit-Explicit (IMEX) splitting method which eliminates the time step restrictions imposed by the diffusion terms. The third order IMEX-Runge-Kutta method improves the time resolution of the method and prevents wrong propagation velocities that can occur if lower order time integration schemes are used, cf. [20]. Further, we interpolate backwards in time to handle the delay term and we introduce a time step adaptation that prevents unnecessary small time steps in the microscopic regime.

Though the techniques which we describe in this section are customized to the model (4.1) they are modular and can be easily adjusted and applied to other chemo-, haptotaxis models which include different time scales, delay terms and non-constant diffusion coefficients. We note however that the splitting of the system as well as the choice of the time step has to be reconsidered in every individual case in order to derive an efficient method.

We consider a two-dimensional computational domain $\Omega = (a,b) \times (a,b) \subset \mathbb{R}^2$, which will be subdivided into a finite number of regular computational cells of size:

$$h = (h_1, h_2)^T \text{ where } h_1 = \frac{b-a}{L}, \; h_2 = \frac{b-a}{M}.$$

Here $L, M \in \mathbb{N}$ denotes the resolution of the grid along the x_1- and x_2-directions, respectively. the total number of grid cells is $N = LM$. The cell centers are located at

$$\mathbf{x}_{1,1} = \left(a + \frac{h_1}{2}\right)\mathbf{e}_1 + \left(a + \frac{h_2}{2}\right)\mathbf{e}_2, \tag{4.8}$$

$$\mathbf{x}_{i,j} = \mathbf{x}_{1,1} + (i-1)\,h_1\,\mathbf{e}_1 + (j-1)\,h_2\,\mathbf{e}_2, \tag{4.9}$$

for $i = 1, \ldots, L, j = 1, \ldots, M$, where $\mathbf{e}_1, \mathbf{e}_2$ are the unit vectors along the x_1- and x_2-directions, respectively. Consequently, the computational cells are given by

$$C_{i,j} = \left\{\mathbf{x}_{i,j} + (\lambda_1 h_1, \lambda_2 h_2),\; \lambda_1, \lambda_2 \in \left[-\frac{1}{2}, \frac{1}{2}\right)\right\},\; i = 1, \ldots, L,\; j = 1, \ldots, M.$$

We introduce a single-index notation for the two-dimensional computational cells using the *lexicographical order*, i.e.

$$C_{i,j} \longrightarrow C_{i+(j-1)L}, \tag{4.10a}$$

$$\mathbf{x}_{i,j} \longrightarrow \mathbf{x}_{i+(j-1)L}, \tag{4.10b}$$

for $i = 1, \ldots, L, j = 1, \ldots, M$, and inversely

$$C_k \longrightarrow C_{k - \lfloor \frac{k-1}{L} \rfloor L, \lfloor \frac{k-1}{L} \rfloor + 1}, \tag{4.11a}$$

$$\mathbf{x}_k \longrightarrow \mathbf{x}_{k - \lfloor \frac{k-1}{L} \rfloor L, \lfloor \frac{k-1}{L} \rfloor + 1}, \tag{4.11b}$$

for $k = 1, \ldots, N$, where $\lfloor \; \rfloor$ is the *Gauss floor function*. We denote moreover by $C_{k \pm \mathbf{e}_j}$ the neighbouring cell of C_k along the positive (negative) $\mathbf{e}_j$ direction ($j = 1, 2$). Hence for $k = 1, \ldots, N$ we have

$$C_{k \pm \mathbf{e}_1} = C_{k - \lfloor \frac{k-1}{L} \rfloor L \pm 1, \lfloor \frac{k-1}{L} \rfloor + 1}, \text{ for } k \neq 0, 1 \bmod L, \text{ respectively}, \tag{4.12a}$$

$$C_{k \pm \mathbf{e}_2} = C_{k - \lfloor \frac{k-1}{L} \rfloor L, \lfloor \frac{k-1}{L} \rfloor + 1 \pm 1}, \text{ for } k \leq L(M-1),\; k \geq L+1, \text{ respectively}. \tag{4.12b}$$

4.3.1 Space Discretization

The system (4.3) is discretised in space by a finite volume method. The approximate solution is represented on every computational cell C_i by a piecewise constant function

$$\mathbf{w}_i(t) \approx \frac{1}{|C_i|} \int_{C_i} \mathbf{w}(x,t)\, dx. \tag{4.13}$$

Moreover, for $\mathbf{w}_h(\cdot) = \{\mathbf{w}_i(\cdot)\}_{i=1}^N$, we consider the following approximations of the advection, diffusion, and reaction operators:

$$\begin{cases} A_i(\mathbf{w}_h(t)) \approx \dfrac{1}{|C_i|} \displaystyle\int_{C_i} A(\mathbf{w}(x,t))\, dx, \\ D_i(\mathbf{w}_h(t)) \approx \dfrac{1}{|C_i|} \displaystyle\int_{C_i} D(\mathbf{w}(x,t))\, dx, \\ R_i(\mathbf{w}_h(t)) \approx \dfrac{1}{|C_i|} \displaystyle\int_{C_i} R(\mathbf{w}(x,t))\, dx. \end{cases} \tag{4.14}$$

Reaction The reaction terms are discretised by a direct evaluation of the reaction operator at the cell centres

$$R_i(\mathbf{w}_h(t)) = R(\mathbf{w}_i(t)). \tag{4.15}$$

Diffusion We denote the discrete diffusion coefficient, see also (4.3), by

$$T_i(\mathbf{w}_h(t)) = \frac{D_c \kappa_i}{1 + (c_{1,i} + c_{2,i})v_i},$$

and define the second component of the discrete diffusion operator using *central differences*

$$\begin{aligned}[D_i(\mathbf{w}_h(t))]_2 = \sum_{j=1}^{2} & \frac{T_{i-\mathbf{e}_j}(\mathbf{w}_h(t)) + T_i(\mathbf{w}_h(t)))}{2 h_j^2} c_{2,i-\mathbf{e}_j} \\ & - \frac{T_{i-\mathbf{e}_j}(\mathbf{w}_h(t)) + 2 T_i(\mathbf{w}_h(t)) + T_{i+\mathbf{e}_j}(\mathbf{w}_h(t))}{2 h_j^2} c_{2,i} \\ & + \frac{T_i(\mathbf{w}_h(t)) + T_{i+\mathbf{e}_j}(\mathbf{w}_h(t))}{2 h_j^2} c_{2,i+\mathbf{e}_j}, \end{aligned} \tag{4.16}$$

with all remaining components $[D_i(\mathbf{w}_h(t))]_j, j = 1, 3, 4, 5$, being equal to zero.

Advection The advection term is discretised using the *central upwind flux*, see [7, 22], which in the particular case of the system (4.3) reads as

$$A_i(\mathbf{w}_h(t)) = \sum_{j=1}^{2} \frac{1}{h_j} \left(0, \ H_{i+\mathbf{e}_j/2}(\mathbf{w}_h(t)) - H_{i-\mathbf{e}_j/2}(\mathbf{w}_h(t)), \ 0, \ 0, \ 0\right)^T. \quad (4.17)$$

The numerical flux $H_{i+\mathbf{e}_j/2}$ approximates the flux between the computational cells C_i and $C_{i+\mathbf{e}_j}$, $j = 1, 2$:

$$H_{i+\mathbf{e}_j/2}(\mathbf{w}_h) = \begin{cases} P_{i+\mathbf{e}_j/2}(\mathbf{w}_h) \, c^{+}_{2,i+\mathbf{e}_j/2}, & \text{if } P_{i+\mathbf{e}_j/2}(\mathbf{w}_h) \geq 0, \\ P_{i+\mathbf{e}_j/2}(\mathbf{w}_h) \, c^{-}_{2,i+\mathbf{e}_j/2}, & \text{if } P_{i+\mathbf{e}_j/2}(\mathbf{w}_h) < 0, \end{cases} \quad (4.18)$$

and $P_{i+\mathbf{e}_j/2}$ represents the local characteristic speeds as:

$$P_{i+\mathbf{e}_j/2}(\mathbf{w}_h) = \frac{D_h}{2} \left(\frac{\kappa_i \, v_i}{1+v_i} + \frac{\kappa_{i+\mathbf{e}_j} \, v_{i+\mathbf{e}_j}}{1+v_{i+\mathbf{e}_j}} \right) \frac{v_{i+\mathbf{e}_j} - v_i}{h_j},$$

for both space directions $j = 1, 2$. The interface values $c^{\pm}_{2,i+\mathbf{e}_j/2}$ are computed by the linear reconstructions

$$c^{-}_{2,i+\mathbf{e}_j/2} = c_{2,i} + s_i^{(j)}, \quad (4.19a)$$

$$c^{+}_{2,i+\mathbf{e}_j/2} = c_{2,i+\mathbf{e}_j} - s_{i+\mathbf{e}_j}^{(j)}, \quad (4.19b)$$

where the slopes $s_i^{(j)}$ are provided by the *monotonized central* (MC) limiter [41]

$$s_i^{(j)} = \text{minmod}\left(c_{2,i} - c_{2,i-\mathbf{e}_j}, \ \frac{1}{4}(c_{2,i+\mathbf{e}_j} - c_{2,i-\mathbf{e}_j}), \ c_{2,i+\mathbf{e}_j} - c_{2,i}\right). \quad (4.20)$$

The *minmod operator* is given by

$$\text{minmod}(v_1, v_2, v_3) = \begin{cases} \max\{v_1, v_2, v_3\}, & \text{if } v_k < 0, \ k = 1, 2, 3, \\ \min\{v_1, v_2, v_3\}, & \text{if } v_k > 0, \ k = 1, 2, 3, \\ 0, & \text{otherwise.} \end{cases} \quad (4.21)$$

Applying the above and (4.14)–(4.17), we obtain the system of the Ordinary Differential Equations (ODEs)

$$\partial_t \mathbf{w}_h - A(\mathbf{w}_h) = R(\mathbf{w}_h) + D(\mathbf{w}_h). \quad (4.22)$$

4.3.2 Time Discretization

We consider $\mathbf{w}_h^n$ a numerical approximation of the solution $\mathbf{w}_h(t)$ of (4.22) at discrete time instances t_n, where $t_n = t_{n-1} + \Delta t_n$. For the choice of the time steps Δt_n we refer to Sect. 4.3.4.

IMEX For the time discretisation of (4.22) we employ an *Implicit-Explicit Runge-Kutta* (IMEX RK) method of 3rd order of accuracy first proposed in [19].

A diagonally implicit RK scheme is applied to the implicit part and an explicit RK scheme to the explicit part. The scheme can be written in the following way:

$$\begin{cases} \mathbf{w}_h^{n+1} = \mathbf{w}_h^n + \Delta t_n \Big(\sum_{j=1}^{i-1} b_j^E (-\mathbf{A} + \mathbf{R}_{\text{expl}})(t_n + c_j^E \Delta t, \mathbf{W}_j) + \sum_{j=1}^{i} b_j^I (\mathbf{D} + \mathbf{R}_{\text{impl}})(\mathbf{W}_j) \Big), \\ \mathbf{W}_i = \mathbf{w}_h^n + \Delta t_n \Big(\sum_{j=1}^{i-1} a_{ij}^E (-\mathbf{A} + \mathbf{R}_{\text{expl}})(t_n + c_j^E \Delta t, \mathbf{W}_j) + \sum_{j=1}^{i} a_{ij}^I (\mathbf{D} + \mathbf{R}_{\text{impl}})(\mathbf{W}_j) \Big), \\ [\mathbf{W}_i]_k = \Big[\mathbf{w}_h^n + \Delta t_n \Big(\sum_{j=1}^{i-1} a_{ij}^E (-\mathbf{A} + \mathbf{R}_{\text{expl}})(t_n + c_j^E \Delta t, \mathbf{W}_j) \Big) \Big]_k, \quad k \in \{1, 3, 5\}. \end{cases}$$
(4.23)

where $\mathbf{b}^E, \mathbf{c}^E \in \mathbb{R}^s$, $A^E \in \mathbb{R}^{s \times s}$, $\mathbf{b}^I, \mathbf{c}^I \in \mathbb{R}^s$ and $A^I \in \mathbb{R}^{s \times s}$ stand for the explicit, and implicit scheme coefficients, respectively. Note that we approximate the advection operator explicitly in time whereas the diffusion terms are treated implicitly. Using the splitting of the reaction operator according to (4.6) and (4.5), the reaction terms are computed both explicitly and implicitly. After the evaluation of the explicit terms we compute the stages $\mathbf{W}_i$ in (4.23) by solving a linear system in the components c_2 and y using the iterative bi-conjugate gradient stabilized *Krylov subspace method* [21, 40].

The particular four stage ($s = 4$) IMEX method that we employ uses the Butcher Table 4.1 and fulfils several stability conditions like A- and L-stability [19].

4.3.3 Treatment of the Delay Term

Of particular importance for the system (4.3) is the *time delay* term $y(t - \chi\tau)$ that appears in the contractivity equation κ. We have included this term in the explicit part $\mathbf{R}_{\text{expl}}$ of the implicit-explicit description (4.6) of R. Consequently, we need to approximate the delay component in the explicit part of the IMEX method.

At stage j of the method (4.23) we evaluate the operator $\mathbf{R}_{\text{expl}}$ at the time instance $\hat{t} = t_n + c_j^E \Delta t_n$, thus we need to approximate $y(\hat{t} - \chi\tau)$. We identify the position of $\hat{t} - \chi\tau$ (recall that $\chi, \tau \geq 0$) and interpolate between the known values of y_h. In some more detail: we consider the time instances $t_1^d \leq t_2^d \leq t_n \leq \hat{t}$, where $t_1^d \leq \hat{t} - \chi\tau$, and corresponding densities of the integrins $y_h(t_1^d)$, $y_h(t_2^d)$, $y_h(t_n)$, $y_h(\hat{t})$. Then the

Table 4.1 Butcher table for the explicit (upper) and the implicit (lower) parts of the third order IMEX scheme (4.23), see also [19]

0				
$\frac{1767732205903}{2027836641118}$	$\frac{1767732205903}{2027836641118}$			
$\frac{3}{5}$	$\frac{5535828885825}{10492691773637}$	$\frac{7880022342437}{10882634858940}$		
1	$\frac{6485989280629}{16251701735622}$	$-\frac{4246266847089}{9704473918619}$	$\frac{10755448449292}{10357097424841}$	
	$\frac{1471266399579}{7840856788654}$	$-\frac{4482444167858}{7529755066697}$	$\frac{11266239266428}{11593286722821}$	$\frac{1767732205903}{4055673282236}$

0	0			
$\frac{1767732205903}{2027836641118}$	$\frac{1767732205903}{4055673282236}$	$\frac{1767732205903}{4055673282236}$		
$\frac{3}{5}$	$\frac{2746238789719}{10658868560708}$	$-\frac{640167445237}{6845629431997}$	$\frac{1767732205903}{4055673282236}$	
1	$\frac{1471266399579}{7840856788654}$	$-\frac{4482444167858}{7529755066697}$	$\frac{11266239266428}{11593286722821}$	$\frac{1767732205903}{4055673282236}$
	$\frac{1471266399579}{7840856788654}$	$-\frac{4482444167858}{7529755066697}$	$\frac{11266239266428}{11593286722821}$	$\frac{1767732205903}{4055673282236}$

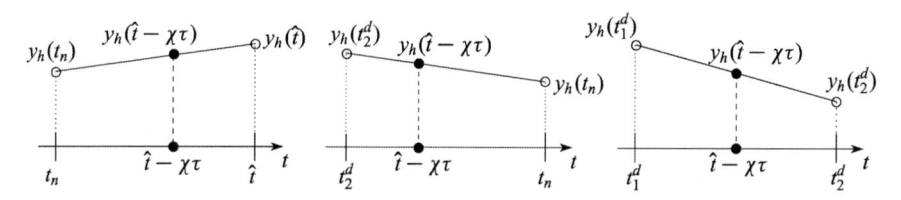

Fig. 4.1 Illustration of the three different cases used to interpolate the delay term

interpolated value of y is given as:

$$y_h(\hat{t} - \chi\tau) = \begin{cases} y_h(t_1^d) + \dfrac{\hat{t} - \chi\tau - t_1^d}{t_2^d - t_1^d}\left(y_h(t_2^d) - y_h(t_1^d)\right), & \text{if } t_1^d \leq \hat{t} - \chi\tau < t_2^d, \\ y_h(t_2^d) + \dfrac{\hat{t} - \chi\tau - t_2^d}{t_n - t_2^d}\left(y_h(t_n) - y_h(t_2^d)\right), & \text{if } t_2^d \leq \hat{t} - \chi\tau < t_n, \\ y_h(t_n) + \dfrac{\hat{t} - \chi\tau - t_n}{\hat{t} - t_n}\left(y_h(\hat{t}) - y_h(t_n)\right), & \text{if } t_n \leq \hat{t} - \chi\tau < \hat{t}, \end{cases}$$

where we use the current numerical solution $y_h(t_n) = [\mathbf{w}_h^n]_4$, and previously computed $y_h(t_1^d), y_h(t_2^d)$. Further, we make the assumption that $W_j \approx w_h(\hat{t})$ and hence employ $y_h(\hat{t}) = [W_j]_4$ to approximate the density of integrins at time instance $\hat{t}$ (Fig. 4.1).

Having computed the time update w_h^{n+1}, we check if the delay time at the next time integration step will overshoot t_2^d. Thus, if $t_{n+1} - \chi\tau \geq t_2^d$ we update the time instances and the corresponding numerical solutions used for the interpolation by

$$t_1^d \leftarrow t_2^d, \quad t_2^d \leftarrow t_n, \quad y_h(t_1^d) \leftarrow y_h(t_2^d), \quad y_h(t_2^d) \leftarrow y_h(t_n).$$

4.3.4 Choice of the Time Step

For stability reasons the time steps are restricted by the characteristic velocities using the CFL condition, [8]:

$$\max_{i=1,\ldots,N,\ j=1,2} \Delta t_n \frac{P_{i+e_j/2}}{h_j} \leq 0.5 . \quad (4.24)$$

Moreover, we note that the ODE subsystem of the last two equations of (4.3) is stiff due to the large parameter $1/\chi$, cf. (4.7). This results in instabilities and inaccuracies in our partly explicit method if the time steps are not further regulated. To cure this problem, we choose Δt_n such that the relative change of κ remains bounded, i.e.

$$\frac{\|\kappa_h(t_n) - \tilde{\kappa}_h(t_n + \Delta t_n)\|_\infty}{\|\kappa_h(t_n)\|_\infty} \leq 0.01 , \quad (4.25)$$

where $\kappa_h(t_n) = [\mathbf{w}_h^n]_5$. Since the time increment Δt_n is needed in our method to compute the actual approximate contractivity $\kappa(t_n + \Delta t_n) = [\mathbf{w}_h^{n+1}]_5$ we apply (4.25) using an estimator $\tilde{\kappa}_h(t_n + \Delta t_n) \approx \kappa_h(t_n + \Delta t_n)$. It can be seen in (4.3) that the component κ evolves quickly (in physical time) to a quasi-steady-state, in order to keep the second order accuracy, cf also [42]. After this state is reached, the changes in κ are very slow. Hence the restriction (4.25) affects the employed time increment only at the beginning of the computation.

The choice of the threshold value 0.01 in (4.25) has followed from numerical experimentation and in order to keep the second order accuracy, see Tables 4.2, 4.3 and Fig. 4.2.

In practice we compute Δt from (4.24) and (4.25) as follows: As a very first step in the IMEX method we compute $A(t_n, \mathbf{w}_h^n)$, $R_{\text{expl}}(t_n, \mathbf{w}_h^n)$ which are needed for the first stage of the RK updates. In the flux computation we get

$$a = \max_{i=1,\ldots,N,\ j=1,2} \frac{P_{i+e_j/2}}{h_j}.$$

Table 4.2 L_1-errors and EOC of the components c_1, c_2, κ

		c_1		c_2		κ	
h	Grid cells	L_1-error	EOC	L_1-error	EOC	L_1-error	EOC
0.08	50×50	2.399e−02		4.195e−02		4.346e−02	
0.04	100×100	6.067e−03	1.9831	1.088e−02	1.9475	1.069e−02	2.0233
0.02	200×200	1.514e−03	2.0026	2.751e−03	1.9831	2.681e−03	1.9956
0.01	400×400	3.785e−04	2.0003	6.912e−04	1.9930	6.724e−04	1.9954

The errors and the EOCs have been computed by (4.27), (4.28) The parameter set and the initial conditions are described in Experiment 4.a (Sect. 4.4.1). See also Fig. 4.2 (left)

Table 4.3 L_2-errors and EOC of the components c_1, c_2, κ

h	Grid cells	c_1 L_2-error	EOC	c_2 L_2-error	EOC	κ L_2-error	EOC
0.08	50×50	5.375e−03		1.303e−02		1.309e−02	
0.04	100×100	1.342e−03	2.0024	3.322e−03	1.9713	3.187e−03	2.0375
0.02	200×200	3.350e−04	2.0018	8.347e−04	1.9928	8.029e−04	1.9890
0.01	400×400	8.369e−05	2.0010	2.090e−04	1.9977	2.012e−04	1.9965

The setting is described in Experiment 4.a

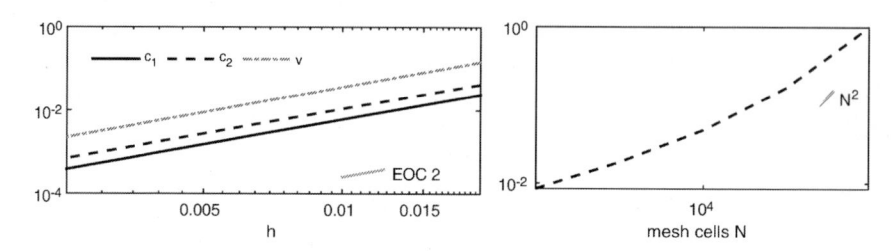

Fig. 4.2 (Left:) Graphical representation of the convergence order for the components c_1, c_2, and v on a two dimensional domain with the grid step size h, see also Tables 4.2, 4.3. (Right:) Graph of the relative computational costs as a function of the total number of computational cells. The results correspond to Experiment 4.a

Since the equation for the contractivity includes only reaction terms that are evaluated in the operator $\mathbf{R}_{\text{expl}}$, we can employ the forward Euler estimator

$$\tilde{\kappa}_h(t_n + \Delta t_n) = \kappa_h(t_n) + \Delta t_n [\mathbf{R}_{\text{expl}}(t_n, \mathbf{W}_i)]_5.$$

We do not actually compute the Euler step $\tilde{\kappa}_h(t_n + \Delta t_n)$, merely substitute in (4.25) and deduce the time increment

$$\Delta t_n = \min\left\{\frac{1}{2a}, \frac{\|\kappa_h(t_n)\|_\infty}{100 \, \|[\mathbf{R}_{\text{expl}}(t_n, \mathbf{W}_i)]_5\|_\infty}\right\}, \quad (4.26)$$

before we compute a new time update. In effect, we compute Δt_n without placing additional computational burden on the method.

This way we resolve efficiently the effect of the microscopic subsystem in the model dynamics instead of choosing small time steps that ensure stability of the explicit method in the stiff contractivity subsystem for the full computation. Our numerical experiments have shown a significant decrease in the CPU usage for the adaptive approach, which yields a speed-up factor of almost five.

4.4 Experimental Results

In Tables 4.2, 4.3 and in Fig. 4.2 (left) we present the *Experimental Order of Convergence* (EOC) rate for the developed method using the parameters and initial data from the Experiment 4.a. The errors are computed by the relative differences between numerical solutions of subsequently refined grids, i.e. for the component k of the system (4.4) we set

$$E_h^k = \|[\mathbf{w}_h^n]_k - [\mathbf{w}_{h/2}^n]_k\|, \quad (4.27)$$

where we use either the discrete L^1 or L^2 norm, and the EOCs are then deduced by the formula

$$\text{EOC}^k = \log_2 E_{h/2}^k - \log_2 E_h^k. \quad (4.28)$$

We can clearly recognize the second order convergence in all five components of the system. The corresponding computational costs are presented in the Fig. 4.2 (right), where the actual values have been scaled with respect to the more expensive grid 400×400.

In Figs. 4.3 and 4.4 we see the initial conditions and the final time solutions of (4.3) according to the particular Experiment 4.b. Despite the smoothness of the initial conditions, a steep front is formed in c_2, see Fig. 4.4 (left). All the components of the solution are presented in Fig. 4.4 (right), where we can also recognize the particular structure of c_2 in detail: a propagating front which seems to be "separated" from the original part of the tumour is followed by a smooth part.

In Fig. 4.5 we present the dependence of the numerical solution of (4.3) on the delay parameter τ. In particular, we recompute Experiment 4.b while varying parameters one-at-a-time and consider the position and height of the steep front of c_2; the aggressiveness of the tumour. We have seen this steep front in Fig. 4.4 and it is present for all tested parameter values. We see that an increasing value of τ leads to a decreasing position of the propagation front of c_2 and an increase of the front height. For values of $\tau \geq 22$ we also see a drop of the front height, which is due

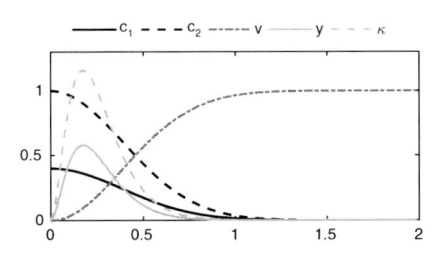

Fig. 4.3 Radial cut (positive semi-axis) of the two-dimensional initial conditions of the Experiment 4.b. See also Fig. 4.4 (right) for the numerical solution at the final time

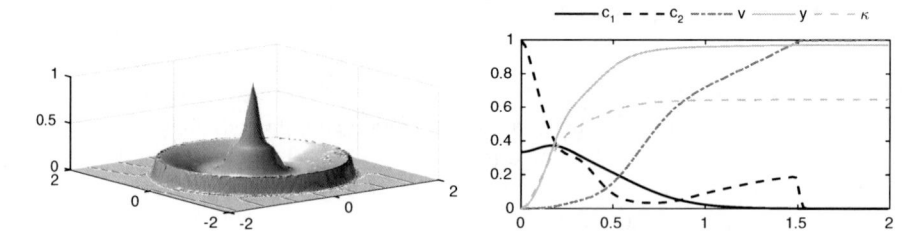

Fig. 4.4 Experiment 4.b. (Left:) Distribution of the migrating c_2 cells for a delay with $\tau = 15$ at the time $t = 0.5$. The invasion pattern exhibits a steep front not existing in the initial condition, cf. Fig. 4.3. (Right:) Radial cuts of all the components of the solution. The component c_2 develops a steep front. The position of the propagating front of c_2 and its magnitude depends on the delay τ, cf. Fig. 4.5

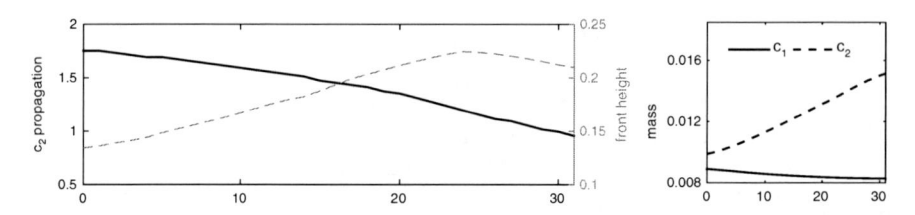

Fig. 4.5 Presented here is the effect of the delay parameter τ in the aggressiveness of the tumour. (Left:) The position of the propagation front of the migrating cells decreases with the increase of the delay (solid line). The height of the propagating front of the c_2 on the other hand increases with the delay up to $\tau \approx 22$ (dashed line). For larger delay value the front height decreases, cf. Fig. 4.6. (Right:) The mass of the migrating cells increases with the delay whereas the mass of the proliferative decreases slightly. The corresponding computational setting is described in Experiment 4.b (Sect. 4.4.1)

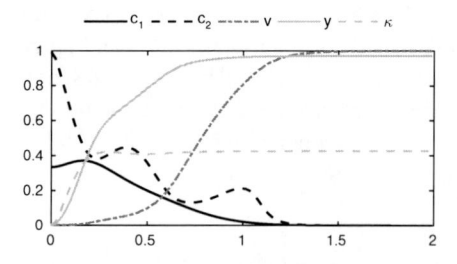

Fig. 4.6 A radial cut of the numerical solution at the final time of the Experiment 4.b for a "large" delay $\tau = 31$. The propagating front has invaded to a lesser extent than for smaller values of τ, cf. Figs. 4.4 (right) and 4.7 (left)

to the emerging of a secondary invasion front, see also Fig. 4.6. We can also see a constant rate increase of the mass of c_2 with τ (right), whereas the mass of c_1 is not significantly influenced.

Moreover, we can also see in Fig. 4.6 that larger delay values τ cause the propagating front of the migratory c_2 cells to invade to a lesser extent than for

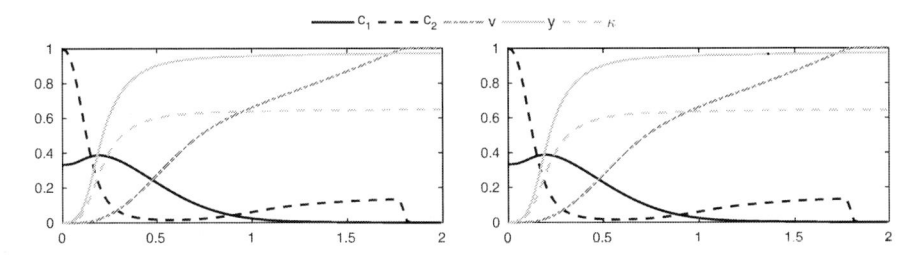

Fig. 4.7 Graphical comparison between $\tau = 1$ and $\chi = 0.001$ (left) versus $\tau = 10$ and $\chi = 0.0001$ (right). The results at the same final time are almost identical. We can so deduce that the "convergence" of the ODE subsystem y, k to the quasi-steady state is very fast, and that the results on the aggressiveness of c_2 is mostly due to the compound delay $\chi\tau$ and less due to the time scale χ. The computational setting is described in Experiment 4.b

smaller τ, cf. Figs. 4.4 and 4.7. A secondary front, that follows closely the primary front, affects its magnitude, see also Fig. 4.4. The experimental setting is given in Experiment 4.b.

In Fig. 4.7 we compare the case $\tau = 1$ and $\chi = 0.001$ (left) versus $\tau = 10$ and $\chi = 0.0001$ (right). We see that the results are almost identical and deduce, since the compound delay $\chi\tau$ is constant, that the rescaling factor χ has small (if any) influence in the dynamics of the system (4.3), despite the stiffness of the subsystem y–κ with respect to χ. We verify this way that the aggressiveness of c_2, as we witness in Fig. 4.5, is mostly influenced by the composite delay $\chi\tau$ and less by the actual time scaling χ.

4.4.1 Description of Experiments

Here we give technical details on the experiments that have been presented in this work. Our simulations have been performed on the computational domain $\Omega = [-2, 2] \times [-2, 2]$. In all experiments we have employed zero Neumann boundary conditions for the advective-diffusive component c_2 of the solution

$$-D_c \frac{\kappa}{1 + (c_1 + c_2)v} \frac{\partial c_2}{\partial \mathbf{n}} + D_h \frac{\kappa v}{1 + v} c_2 \frac{\partial v}{\partial \mathbf{n}} = 0$$

where $\mathbf{n}$ is the outward normal vector to the computational domain Ω.

Experiment 4.a This experiment corresponds to the convergence results in Fig. 4.2 and Tables 4.2, 4.3. Following the original derivation of the model [36], we consider the following set of parameters:

$$\begin{cases} \mu_c = 1, & \eta_1 = 0.05, & \gamma = 0.055, & \lambda = 0.076, \\ D_c = 10^{-3}, & D_h = 1, \\ \delta_v = 10, & \mu_v = 0.3, & \eta_2 = 0.9, \\ k_1 = 2, & k_{-1} = 0.06, \\ q = 3, & M = 1, \\ \chi = 0.01, & \tau = 0.04. \end{cases} \quad (4.29)$$

The initial condition reads

$$\begin{cases} c_1(0, x) = 0.4\, e^{-\frac{1}{\varepsilon}(x^2+y^2)}, \\ c_2(0, x) = e^{-\frac{1}{\varepsilon}(x^2+y^2)}, \\ v(0, x) = 1 - c_2(0, x), \\ y(0, x) = 20 f_\gamma \left(5\left(x^2 + y^2\right), 2, 15\right), \\ \kappa(0, x) = 2 y(0, x), \end{cases} \quad (4.30)$$

for $\varepsilon = 1.5$ and $x \in \Omega$, where we employ the density function of the gamma distribution,

$$f_\gamma(x, a, b) = \frac{1}{b^a \Gamma(a)} x^{a-1} e^{\frac{-x}{b}}, \quad \text{where } \Gamma(a) = \int_0^\infty t^{a-1} e^{-t} dt. \quad (4.31)$$

These initial conditions are inspired by Meral et al. [24], where the authors justified their choice based on the expected spatial monotonicity of the ligand distribution.

Experiment 4.b The parameters and initial conditions that follow, correspond to Figs. 4.3, 4.4, 4.5, 4.6, 4.7. Parameters

$$\begin{cases} \mu_c = 1, & \eta_1 = 0.05, & \gamma = 0.055, & \lambda = 0.152, \\ D_c = 10^{-2}, & D_h = 10, \\ \delta_v = 5, & \mu_v = 0.3, & \eta_2 = 0.9, \\ k_1 = 2, & k_{-1} = 0.06, \\ q = 3, & M = 1, \\ \chi = 0.01, & \tau = 0.04. \end{cases} \quad (4.32)$$

Initial conditions

$$\begin{cases} c_1(0,x) = 0.4\, e^{-\frac{1}{\varepsilon}(x^2+y^2)}, \\ c_2(0,x) = e^{-\frac{1}{\varepsilon}(x^2+y^2)}, \\ v(0,x) = 1 - c_2(0,x), \\ y(0,x) = 15 f_\gamma\left(80\sqrt{x^2+y^2}, 3, 7\right), \\ \kappa(0,x) = 2\, y(0,x), \end{cases} \quad (4.33)$$

where $\varepsilon = 1.5$, $x \in \Omega$, and f_γ is defined in (4.31).

4.5 Conclusions

Since their first derivation, cancer growth models have been a theatre for the development of new problem-suited numerical methods. This is not only due to the importance of the topic, but primarily, to the complex dynamics of the solutions. Our work aims to be a contribution along these lines.

We solve numerically the model (4.3) that was proposed in [36]. The method we employ is a concatenation of a robust, positivity preserving FV method in space with a third order IMEX method in time. The additional challenges that we have addressed are the non-constant diffusion coefficients in c_2, the time delay in κ, as well as the stiff reaction terms in y and κ.

We have discretised the non-constant diffusion coefficient using central differences and solved the (implicit) linear system by the Krylov subspace method. For the delay term we perform an interpolation in time between a small number of previously "saved" time steps. We treat the stiffness applying a secondary condition (besides the CFL) by adjusting the time step of the method. The additional condition leads to an adaptive time stepping by employing an explicit Euler step of the κ equation.

We verify numerically that our method is second order accurate and we identify its computational cost. Our numerical experiments indicate that the migrating cancer cells develop a steep propagating front. We also see that the aggressiveness of the tumour depends on the time delay.

References

1. Alt W, Lauffenburger D (1987) Transient behavior of a chemotaxis system modelling certain types of tissue inflammation. J Math Biol 24(6):691–722
2. Andasari V, Gerisch A, Lolas G, South A, Chaplain M (2011) Mathematical modelling of cancer cell invasion of tissue: biological insight from mathematical analysis and computational simulation. J Math Biol 63(1):141–171

3. Anderson A, Chaplain M, Newman E, Steele R, Thompson A (2000) Mathematical modelling of tumour invasion and metastasis. Comput Math Methods Med 2(2):129–154
4. Armitage P, Doll R (1954) The age distribution of cancer and a multi-stage theory of carcinogenesis. Br J Cancer 8(1):1
5. Bellomo N, Li N, Maini P (2008) On the foundations of cancer modelling: selected topics, speculations, and perspectives. Math Models Methods Appl Sci 18(04):593–646
6. Chaplain M, Lolas G (2005) Mathematical modelling of cancer cell invasion of tissue. the role of the urokinase plasminogen activation system. Math Models Methods Appl Sci 15(11):1685–1734
7. Chertock A, Kurganov A (2008) A second-order positivity preserving central-upwind scheme for chemotaxis and haptotaxis models. Numer Math 111(2):169–205
8. Courant R, Friedrichs K, Lewy H (1928) über die partiellen differenzengleichungen der mathematischen physik. Math Ann 100(1):32–74
9. Domschke P, Trucu D, Gerisch A, Chaplain M (2014) Mathematical modelling of cancer invasion: Implications of cell adhesion variability for tumour infiltrative growth patterns. J Theor Biol 361:41–60
10. Fisher J (1958) Multiple-mutation theory of carcinogenesis. Nature 181(4609):651–652
11. Ganguly R, Puri I (2006) Mathematical model for the cancer stem cell hypothesis. Cell Prolif 39(1):3–14
12. Gao D, Vahdat L, Wong S, Chang J, Mittal V (2012) Microenvironmental regulation of epithelial-mesenchymal transitions in cancer. Cancer Res 72(19):4883–4889
13. Gerisch A, Chaplain M (2008) Mathematical modelling of cancer cell invasion of tissue: local and nonlocal models and the effect of adhesion. J Theor Biol 250(4):684–704
14. Gupta P, Chaffer C, Weinberg R (2009) Cancer stem cells: mirage or reality? Nat Med 15(9):1010–1012
15. Hellmann N, Kolbe N, Sfakianakis N (2016) A mathematical insight in the epithelial-mesenchymal-like transition in cancer cells and its effect in the invasion of the extracellular matrix. Bull Braz Math Soc 47(1):397–412
16. Johnston M, Maini P, Jonathan-Chapman S, Edwards C, Bodmer W (2010) On the proportion of cancer stem cells in a tumour. J Theor Biol 266(4):708–711
17. Katsuno Y, Lamouille S, Derynck R (2013) TGF-β signaling and epithelial–mesenchymal transition in cancer progression. Curr Opin Oncol 25(1):76–84
18. Keller E, Segel L (1970) Initiation of slime mold aggregation viewed as an instability. J Theor Biol 26(3):399–415
19. Kennedy C, Carpenter M (2003) Additive Runge-Kutta schemes for convection-diffusion-reaction equations. Appl Numer Math 1(44):139–181
20. Kolbe N, Kať uchová J, Sfakianakis N, Hellmann N, Lukáčová-Medvid'ová M (2016) A study on time discretization and adaptive mesh refinement methods for the simulation of cancer invasion: the urokinase model. Appl Math Comput 273:353–376
21. Krylov A (1931) On the numerical solution of the equation by which in technical questions frequencies of small oscillations of material systems are determined. Otdel mat i estest nauk VII(4):491–539
22. Kurganov A, Lukáčová-Medvid'ová M (2014) Numerical study of two-species chemotaxis models. Discrete Cont Dyn-B 19(1):131–152
23. Mani S, Guo W, Liao M, Eaton E, Ayyanan A, Zhou A, Brooks M, Reinhard F, Zhang C, Shipitsin M, Campbell L, Polyak K, Brisken C, Yang J, Weinberg R (2008) The epithelial-mesenchymal transition generates cells with properties of stem cells. Cell 133(4):704–715
24. Meral G, Stinner C, Surulescu C (2015) On the multiscale model involving cell contractivity and its effects on tumour invasion. Discret Cont Dyn Syst 20:189–213
25. Michor F (2008) Mathematical models of cancer stem cells. J Clin Oncol 26(17):2854–2861
26. Neagu A, Mironov V, Kosztin I, Barz B, Neagu M, Moreno-Rodriguez R, Markwald R, Forgacs G (2010) Computational modelling of epithelial–mesenchymal transformations. Biosystems 100(1):23–30
27. Nordling C (1953) A new theory on the cancer-inducing mechanism. Br J Cancer 7(1):68

28. Painter K, Hillen T (2011) Spatio-temporal chaos in a chemotaxis model. Phys D 240(4):363–375
29. Patlak C (1953) Random walk with persistence and external bias. Bull Math Biophys 15:311–338
30. Perumpanani A, Sherratt J, Norbury J, Byrne H (1996) Biological inferences from a mathematical model for malignant invasion. Invasion Metastasis 16(4–5):209–221
31. Preziosi L (2003) Cancer modelling and simulation. CRC, Boca Raton
32. Reya T, Morrison S, Clarke M, Weissman I (2001) Stem cells, cancer, and cancer stem cells. Nature 414(6859):105–111
33. Sfakianakis N, Kolbe N, Hellmann N, Lukacova M (2016) A multiscale approach to the migration of cancer stem cells: mathematical modelling and simulations. arXiv: 160405056
34. Singh A, Settleman J (2010) EMT, cancer stem cells and drug resistance: an emerging axis of evil in the war on cancer. Oncogene 29(34):4741–4751
35. Stiehl T, Marciniak-Czochra A (2012) Mathematical modeling of leukemogenesis and cancer stem cell dynamics. Math Model Nat Phenom 7(01):166–202
36. Stinner C, Surulescu C, Uatay A (2015) Global existence for a go-or-grow multiscale model for tumor invasion with therapy. Preprint. http://nbn-resolving.de/urn/resolver.pl?urn:nbn:de:hbz:386-kluedo-42943
37. Szymanska Z, Rodrigo C, Lachowicz M, Chaplain M (2009) Mathematical modelling of cancer invasion of tissue: the role and effect of nonlocal interactions. Math Models Methods Appl Sci 19(02):257–281
38. Thiery J (2002) Epithelial–mesenchymal transitions in tumour progression. Nat Rev Cancer 2(6):442–454
39. Vainstein V, Kirnasovsky O, Kogan Y, Agur Z (2012) Strategies for cancer stem cell elimination: insights from mathematical modelling. J Theor Biol 298:32–41
40. van der Vorst HA (1992) Bi-CGSTAB: a fast and smoothly converging variant of Bi-CG for the solution of nonsymmetric linear systems. SIAM J Sci Comput 13(2):631–644
41. Van Leer B (1977) Towards the ultimate conservative difference scheme. IV. A new approach to numerical convection. J Comput Phys 23(3):276–299
42. Wiebe B (2016) Numerical simulations of multiscale cancer invasion models. Master's thesis, University of Mainz. Supervised by M. Lukáčová-Medvid'ová, N. Sfakianakis

Chapter 5
Modelling Tissue Self-Organization: From Micro to Macro Models

Pierre Degond and Diane Peurichard

5.1 Introduction

Self-organization in biological systems is a process that occurs over time and leads to the spontaneous emergence of spatio-temporal structures as a result of simple interactions between agents [6]. The evolution and development of biological self-organization of systems proceeds from small, simple components that are assembled together to form larger structures that have emergent properties and behaviour, which, in turn, self-assemble into more complex structures. The latter are then maintained by a permanent turnover of different cell populations or components, which ensures the proper functioning of the tissue. Multiple evidence show that disrupting this homeostasis promotes tissue dysfunctions and diseases including cancer. Similar unbalance is a natural consequence of aging, which becomes a major problem worldwide. Understanding what factors are responsible for tissue homeostasis disruption is therefore a major issue in the science of aging. Because self-organization in biological systems involves several agents and interactions at different scales and of several types (chemical, mechanical, molecular, genomics), identifying which mechanisms are primarily involved in this organization is of

P. Degond
Imperial College London, London SW7 2AZ, UK
e-mail: pdegond@imperial.ac.uk

D. Peurichard (✉)
MAMBA-Modelling and Analysis for Medical and Biological Applications, LJLL-Laboratoire Jacques-Louis Lions, INRIA de Paris, Université Pierre et Marie Curie, 75252 Paris 05, France
e-mail: diane.a.peurichard@inria.fr

© Springer International Publishing AG, part of Springer Nature 2017
A. Gerisch et al. (eds.), *Multiscale Models in Mechano and Tumor Biology*, Lecture Notes in Computational Science and Engineering 122, https://doi.org/10.1007/978-3-319-73371-5_5

tremendous difficulty. Mathematical models then provide a way of reducing the complexity of the problem, by featuring a finite set of agents and interactions that are supposed to contribute the most to the global organization of the system.

Due to their simplicity and flexibility, the most used models in the literature are Individual-Based Models (IBM), which describe the motion of each individual [1, 3, 10, 15]. An other advantage of these models is that they can incorporate any number of individual-level mechanisms. As a drawback, they are not suited to the study of large systems since the computational cost of an IBM tremendously increases with the size of the system.

The behavior of a large system of individuals can be studied through mesoscopic descriptions based on the evolution of the probability density of finding individuals in the phase space. These descriptions are usually expressed in terms of kinetic partial differential equations obtained by scaling (mean-field) limit of an IBM [16]. Finally, continuum models are proposed to describe the system at the macroscopic level. These last models describe the evolution in time of mean variables such as density, mean orientation etc. The main advantage of such models is their low computational cost. As a drawback, information on the interactions are lost at the individual level. To overcome this weakness of the macroscopic models, a possible route is to derive a macroscopic model from a microscopic one, topic which has received a lot of attention these last decades [5, 8, 11, 14]. The classical derivation consists of two steps: (1) obtain a kinetic version of the microscopic model by asymptotic limit of large number of individuals and (2) perform the large scale limit of the kinetic model to obtain a macroscopic description.

In this chapter, we aim to apply these recent techniques to study the formation of specific structures in biological tissues. Numerous models of tissue morphogenesis can be found in the literature, describing the emergence of self-organization of cells and fibers (see [17, 23] and references therein). Biological fiber networks alone have also been extensively studied in the literature. These complex networks are cross-linked dynamical plastic structures providing mechanical support to the cells and giving to the tissue the ability to change shape and adapt in response to biological and mechanical stimuli [7]. At the macroscopic level, numerous continuum models for fibrous media have also been developed in the literature. Most of them are heuristically derived from continuum theories such as [2, 22], thermodynamics [18], or viscous fluid mechanics [19]. The challenge for these models is to construct accurate constitutive laws and homogenization techniques to incorporate the dynamics of the fiber network, even if it implies a loss of information at the individual level.

The goal of this chapter is to summarize some recent works aiming at deriving a macroscopic model from an agent-based model for interconnected fibers interacting through alignment interactions first proposed in [9]. We start from an agent-based model for fibers having the ability to cross-link and unlink. Two linked fibers interact through alignment interaction at the cross link. In the asymptotic limit of a large number of individuals, we obtain a closed system of two equations describing the time evolution of the one-particle fiber distribution function and of the cross-links distribution function. The cross-links provide correlations between the fibers and consequently their distribution can be viewed as related to the two-particle fiber distribution. It is shown that the knowledge of the one-particle distributions of fibers and of cross links suffices to provide a physically relevant kinetic description of the system.

We then derive a macroscopic model by exploring the diffusion limit of the kinetic model with further scaling assumptions on the model parameters. In order to obtain a closure of the kinetic model at the level of the fiber distribution function only, we suppose that the linking/unlinking frequencies are very large: the typical linking/unlinking time is supposed to be much smaller than the typical fiber alignment time. The biological relevance of this assumption may be questioned, our main goal here is to set up a methodology which will be further refined towards providing an extensive theory of fibrous media with finite linking/unlinking times.

From these assumptions, the rescaled kinetic problem has the form of a classical diffusion approximation problem whose leading-order collision operator comes from the nematic alignment of the fibers due to the cross-links. This operator has equilibria in the form of generalized von Mises distributions of the fiber directions. Such equilibria are also observed in other systems featuring polar or nematic alignment between particles [4, 12, 13]. Therefore, the fiber distribution function is described at the macroscopic level by a fiber density and mean local orientation, raising the need for two equations providing the spatio-temporal evolution of these parameters. As there is no conservation equation other than mass conservation in the model, the concept of Generalized Collision Invariants [8] is used to obtain the macroscopic system of equations for the fiber density and mean orientation. In the case of a homogeneous fiber distribution, when the density is uniform in space and constant in time, the resulting macroscopic model consists of a quasilinear parabolic equation for the fiber local mean orientation.

5.2 Individual Based Model for Fibers Interacting Through Alignment Interactions

In this section, we sketch the two-dimensional agent-based model for interconnected fibers interacting through alignment interactions at the cross-links proposed in [9]. In this model, long collagen fibers are modelled as sets of segments of uniform fixed length having the ability to connect and disconnect with their intersecting neighbors. In this way, several sequentially cross-linked fiber elements model a long fiber having the ability to bend or even take possible tortuous geometries. Moreover, a link between two connected segments can be positioned at any point along this element (not only the extremities) and a given segment can be connected to any number of other segments, thereby allowing to model the branching off of a fiber into several branches. The topology of the fiber network is constantly remodelled through link creation/deletion processes, following (random) Poisson processes in time. The cross-linking process models fiber elongation and symmetrically, spontaneous unlinking of cross-linked fibers accounts for fiber breakage describing extra-cellular matrix (ECM) remodelling processes. We consider the following phenomena: (1) all along the link lifetime, cross-linked fibers are forced to stay linked by means of a restoring force. (2) To model fiber resistance to bending, we suppose that pairs of linked fibers are subject to a torque that tends to align the two fibers with respect to each other. Finally, (3) the fibers are subject to random positional and orientational noises to model the movements of the tissue and (4) to positional and orientational potential forces, to model the action of external elements (such as cells or other tissues).

We consider a set of N fiber segments of uniform and fixed length L, each described by its center $X_i \in \mathbb{R}^2$ and its angle $\theta_i \in [-\frac{\pi}{2}, \frac{\pi}{2})$ modulo π with respect to a fixed reference direction. Each of the previously described phenomena (1)–(4) is related to an energy functional, namely the energy for the maintenance of the links W_{links}, the energy for the alignment torque W_{align}, the energy for the noise contribution W_{noise} and the energy for the action of the external elements W_{ext}. The total energy is then defined as the sum of all these energies:

$$W_{\text{tot}} = W_{\text{links}} + W_{\text{align}} + W_{\text{noise}} + W_{\text{ext}}.$$

All these energies are functions of the N fiber positions $(X_i)_{i=1}^N$ and orientations $(\theta_i)_{i=1}^N$. Since W_{noise} is rather an entropy than an energy, W_{tot} is indeed the total free energy of the system. Fiber motion and rotation during a time interval between two fiber linking/unlinking events are supposed to follow the steepest descent direction of the total free energy, according to:

$$\frac{dX_i}{dt} = -\mu \nabla_{X_i} W_{\text{tot}}, \quad \forall i \in \{1, \ldots, N\}, \tag{5.1}$$

$$\frac{d\theta_i}{dt} = -\lambda \partial_{\theta_i} W_{\text{tot}}, \quad \forall i \in \{1, \ldots, N\}. \tag{5.2}$$

Equations (5.1) and (5.2) express the motion and rotation of the individuals in an overdamped regime in which the forces due to friction are very large compared to the inertial forces. Fiber velocity and angular speed are proportional to the force exerted on the fiber through two mobility coefficients μ and λ which are considered given.

We recall that link creation and suppresion are supposed to follow Poisson processes of frequencies ν_f and ν_d respectively. The probability that a link is created (resp. deleted) in the time interval $[t_k, t]$ is $1 - e^{-\nu_f(t-t_k)}$ (resp. $1 - e^{-\nu_d(t-t_k)}$). Given a time t at which no linking/unlinking process occurs, the set of cross-links between fibers is well defined and **supposed to have K elements**. The energies W_{links} and W_{align} of the total free energy of the system are supposed to be the sums of elementary binary potential elements computed between pairs of linked fibers,

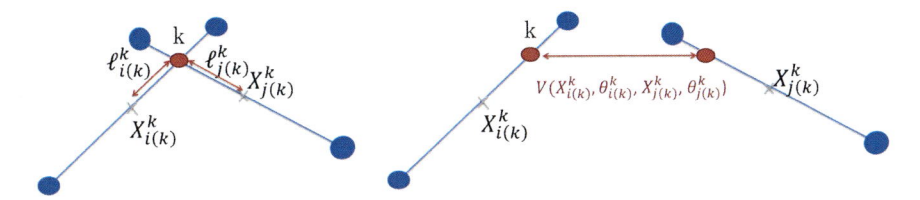

Fig. 5.1 Left: Link k connecting fibers $i(k)$ and $j(k)$. The associated link lengths $\ell_{i(k)}$ and $\ell_{j(k)}$ are indicated on the Figure. Right: restoring potential V between linked fibers $i(k)$ and $j(k)$

ranging over the K links of the system at time t:

$$W_{\text{links}} = \sum_{k=1}^{K} V(X_{i(k)}, \theta_{i(k)}, X_{j(k)}, \theta_{j(k)}) \tag{5.3}$$

$$W_{\text{align}} = \sum_{k=1}^{K} b(\theta_{i(k)}, \theta_{j(k)}) = \sum_{k=1}^{K} \alpha \sin^2(\theta_{i(k)} - \theta_{j(k)}). \tag{5.4}$$

The potential element V relating two linked fibers numbered $i(k)$ and $j(k)$ at their junction k is supposed to derive from a spring-like force that attracts the attachment sites of the two fibers as soon as they are displaced with respect to each other (see Fig. 5.1). The corresponding binary potential element V depends on the positions $X_{i(k)}$, $X_{j(k)}$ and orientations $\theta_{i(k)}$, $\theta_{j(k)}$ of the two linked fibers, as well as on the attachment sites $X_{i(k)}^k$, $X_{j(k)}^k$ defined by:

$$X_{i(k)}^k = X_{i(k)} + \ell_{i(k)}^k \omega_{i(k)}, \quad X_{j(k)}^k = X_{j(k)} + \ell_{j(k)}^k \omega_{j(k)},$$

where $\omega_i = \omega(\theta_i) = \begin{pmatrix} \cos \theta_i \\ \sin \theta_i \end{pmatrix}$ is the unit directional vector of fiber i, $\ell_{i(k)}^k \in [-L/2, L/2]$ (resp. $\ell_{j(k)}^k$) is the algebraic distance of the attachment site of fiber $i(k)$ (resp. $j(k)$) to its center, at the time of creation of the link. We stress out the fact that the quantities $\ell_{i(k)}^k$ and $\ell_{j(k)}^k$ remain constant throughout the link lifetime.

The linked fiber-fiber alignment potential element b is supposed to be proportional to the square of the sine of the angle between two linked fibers $i(k), j(k)$ and only depends on the orientational angles $\theta_{i(k)}, \theta_{j(k)}$.

The external potential W_{ext} associated with the external forces is supposed to be the sum of potential forces $U(X_i, \theta_i)$ acting on each of the N fibers:

$$W_{\text{ext}} = \sum_{i=1}^{N} U(X_i, \theta_i).$$

In the case where the system describes the collagen fibers in a tissue, U aims to model the presence of cells or other organs.

We include random positional and orientational motion of the fiber elements which, in the context of biological tissues, originate from the random movements of the subject. With this aim, we introduce an entropy term:

$$W_{\text{noise}} = d \sum_{i=1}^{N} \log(\tilde{f})(X_i, \theta_i), \tag{5.5}$$

where $\tilde{f}$ is a 'regularized density' describing fibers located around point X_i and of orientation around θ_i (see [9] for details). Such an entropy term gives rise to diffusion terms at the level of the mean-field kinetic model.

In [20], simulations of this 2D microscopic model have been performed on a square domain with periodic boundary conditions, and it has been shown that Eqs. (5.1)–(5.2) correspond to a gradient descent for a quadratic penalization of a minimization problem related to the model of [21]. This last model has proved its efficiency in the modeling of complex cross-linked structures such as the ECM of adipose tissues. The study was supported by quantitative comparisons between numerical simulations and images acquired from biological experiments. From the physical point of view, phase transitions have been shown to take place when some of the model parameters are varied, namely the fiber linking/unlinking frequencies and the cross-linked fiber alignment force. However, this model was shown to be very time-consuming for simulations at the scale of the whole tissue, raising the need for the formulation of a macroscopic level.

In the next sections, we sketch the recent results of [9] which represent a first step towards the derivation of a macroscopic model for interconnected fiber networks from its underlying microscopic model. The formal derivation consists of two main steps: (1) derivation of a kinetic model from the microscopic formulation, in the limit of a large number of fibers and cross-links and (2) a diffusion limit of the kinetic model under appropriate scaling assumptions to obtain the macroscopic model.

5.3 Derivation of a Kinetic Model

In order to obtain a kinetic description of the previously described microscopic model, the empirical measure of the individual fibers, $f^N(x, \theta, t)$, is introduced:

$$f^N(x, \theta, t) = \frac{1}{N} \sum_{i=1}^{N} \delta_{(X_i(t), \theta_i(t))}(x, \theta),$$

where $\delta_{(X_i(t), \theta_i(t))}(x, \theta)$ is the Dirac delta located at $(X_i(t), \theta_i(t))$. Classically, $f^N(x, \theta, t)$ gives the probability to find a fiber within a volume dx about x with orientational angle within $d\theta$ about θ at time t. The empirical measure

$g^K(x_1, \theta_1, \ell_1, x_2, \theta_2, \ell_2, t)$ of the fiber links is given by:

$$g^K = \frac{1}{2K} \sum_{k=1}^{K} \delta_{(X_{i(k)}, \theta_{i(k)}, \ell_{i(k)}^k, X_{j(k)}, \theta_{j(k)}, \ell_{j(k)}^k)}(x_1, \theta_1, \ell_1, x_2, \theta_2, \ell_2)$$
$$+ \delta_{(X_{j(k)}, \theta_{j(k)}, \ell_{j(k)}^k, X_{i(k)}, \theta_{i(k)}, \ell_{i(k)}^k)}(x_1, \theta_1, \ell_1, x_2, \theta_2, \ell_2),$$

with a similar definition of the Dirac deltas. It gives the probability of finding a link with associated lengths within a volume $d\ell_1 d\ell_2$ about ℓ_1 and ℓ_2, this link connecting a fiber located within a volume $dx_1 \frac{d\theta_1}{\pi}$ about (x_1, θ_1) with a fiber located within a volume $dx_2 \frac{d\theta_2}{\pi}$ about (x_2, θ_2). One notes that (ℓ_1, ℓ_2) is defined in $[-\frac{L}{2}, \frac{L}{2}]^2$. Then, at the limit $N, K \to \infty$, $\frac{K}{N} \to \xi$, where $\xi > 0$ is a fixed parameter, $f^N \to f$, $g^K \to g$ where f and g satisfy the equations given in the following theorem (see [20]):

Theorem 5.1 *The formal limit of Eqs. (5.1), (5.2) for $K, N \to \infty$, $\frac{K}{N} \to \xi$, where $\xi > 0$ is a fixed parameter reads:*

$$\frac{\partial f}{\partial t} - \mu \left(\nabla_x \cdot ((\nabla_x U) f) + \xi \nabla_x \cdot F_1 + d \Delta_x f \right) - \lambda \left(\partial_\theta ((\partial_\theta U) f) + \xi \partial_\theta F_2 + d \partial_{\theta}^2 f \right) = 0, \quad (5.6)$$

and

$$\begin{aligned}\frac{\partial g}{\partial t} &- \mu \Big(\nabla_{x_1} \cdot \big(g \nabla_x U(x_1, \theta_1) + \xi \frac{g}{f(x_1, \theta_1)} F_1(x_1, \theta_1) \big) \\
&+ \nabla_{x_2} \cdot \big(g \nabla_x U(x_2, \theta_2) + \xi \frac{g}{f(x_2, \theta_2)} F_1(x_2, \theta_2) \big) \\
&+ d \nabla_{x_1} \cdot \big(\frac{g}{f(x_1, \theta_1)} \nabla_x f(x_1, \theta_1) \big) + d \nabla_{x_2} \cdot \big(\frac{g}{f(x_2, \theta_2)} \nabla_x f(x_2, \theta_2) \big) \Big) \\
&- \lambda \Big(\partial_{\theta_1} \big(g \partial_\theta U(x_1, \theta_1) + \xi \frac{g}{f(x_1, \theta_1)} F_2(x_1, \theta_1) \big) \\
&+ \partial_{\theta_2} \big(g \partial_\theta U(x_2, \theta_2) + \xi \frac{g}{f(x_2, \theta_2)} F_2(x_2, \theta_2) \big) \\
&+ d \partial_{\theta_1} \big(\frac{g}{f(x_1, \theta_1)} \partial_\theta f(x_1, \theta_1) \big) + d \partial_{\theta_2} \big(\frac{g}{f(x_2, \theta_2)} \partial_\theta f(x_2, \theta_2) \big) \Big) = S(g),\end{aligned} \quad (5.7)$$

where

$$F_1(x_1, \theta_1) = \int (g \nabla_{x_1} V)(x_1, \theta_1, \ell_1, x_2, \theta_2, \ell_2) d\ell_1 d\ell_2 \frac{d\theta_2}{\pi} dx_2,$$

$$F_2(x_1, \theta_1) = \int (g(\partial_{\theta_1} V + \partial_{\theta_1} b))(x_1, \theta_1, \ell_1, x_2, \theta_2, \ell_2) d\ell_1 d\ell_2 \frac{d\theta_2}{\pi} dx_2,$$

and $S(g)$ is given by:

$$S(g) = \nu_f f(x_1, \theta_1) f(x_2, \theta_2) \delta_{\bar{\ell}(x_1,\theta_1,x_2,\theta_2)}(\ell_1) \delta_{\bar{\ell}(x_2,\theta_2,x_1,\theta_1)}(\ell_2) - \nu_d g, \qquad (5.8)$$

where $\delta_{\bar{\ell}}(\ell_1)$ denotes the Dirac delta at $\bar{\ell}$, i.e. the distribution acting on test functions $\phi(\ell_1)$ such that $\langle \delta_{\bar{\ell}}(\ell_1), \phi(\ell_1) \rangle = \phi(\bar{\ell})$.

This kinetic model consists of two evolution equations. Equation (5.6) is an equation for the individual fibers and describes the evolution of the one-particle distribution function f. Equation (5.7) is an equation for the links between fiber pairs, where g describes the cross-link distribution function. It can be related to the two-particle fiber distribution function. As the links are tightly tied to the fibers, they are convected by them and follow their motion. Simultaneously, they constrain the linked fibers to move together, so they directly influence their motion. The action of the links on the individual fiber motion is contained in the force terms F_1 and F_2 of Eq. (5.6) where V is the restoring potential which forces cross-linked fibers to stay connected (contained in the term W_{links} of the microscopic model). The second and fifth terms of Eq. (5.6) describe transport in physical and orientational spaces due to the external potential U (contained in the potential W_{ext} of the microscopic model). The kinetic counterpart of the alignment force between linked fibers (contained in the term W_{align} of the microscopic model) is encompassed in the second term of the force F_2 and only acts on the orientation of the fibers. The fourth and seventh terms of Eq. (5.6) are diffusion terms of amplitude λd and μd respectively. They represent the random motion of the fibers and originate from the interactions described in W_{noise} of the microscopic model. The individual motion of the fibers is thus related to the motion of their linked neighbors. The left-hand side of Eq. (5.7) describes the evolution of the links between fibers. It is composed of the convective terms generated by the external potential and by the diffusion terms. The forces induced by the restoring potential generated by the links again give rise to the nonlocal terms F_1 and the first term of F_2. The right hand side $S(g)$ of Eq. (5.7) results from the Poisson processes of linking/unlinking intersecting fibers at frequencies ν_f and ν_d, respectively. The first term of Eq. (5.8) describes the formation of the link and the Dirac deltas indicate that, at the link creation time, the link lengths ℓ_1 and ℓ_2 (in $[-L/2, L/2]$) are set by the geometric configuration of the intersecting fibers at the attachment time. The second term describes fiber unlinking at the rate set by the Poisson process, i.e. ν_d.

The rigorous proof of the convergence of the microscopic model to the kinetic model is still an open question, and this derivation still needs validation through theoretical analysis and numerical simulations. However, this model is to our knowledge a unique explicit example of a kinetic model written in terms of the one and two particle distribution functions and closed at this level. Moreover, the distribution function g can be seen as a way of describing the random graph of the fiber links, a description which could be useful to describe other kinds of random networks. In the sequel, the diffusion limit of this kinetic model is performed via rescaling of space and time, and introducing scaling assumptions on the model parameters.

5.4 Scaling and Macroscopic Model

In order to perform a diffusion limit of the kinetic model of the previous section, we introduce a small parameter $\varepsilon \ll 1$ and set $\tilde{x} = \sqrt{\varepsilon} x$ and $\tilde{t} = \varepsilon t$. This leads to $\tilde{\ell} = \sqrt{\varepsilon}\ell$, $\tilde{f}(\tilde{x},\theta) = \varepsilon^{-1} f(x,\theta)$ and $\tilde{g}(\tilde{x}_1,\theta_1,\tilde{\ell}_1,\tilde{x}_2,\theta_2,\tilde{\ell}_2) = \varepsilon^{-3} g(x_1,\theta_1,\ell_1,x_2,\theta_2,\ell_2)$. We then introduce the following scaling hypothesis on the model parameters: We suppose that the external potential $U(x,\theta)$ is decomposed into $U(x,\theta) = U^0(x) + U^1(\theta)$, where U^0 is acting on the space variable only and U^1 is a π-periodic potential only acting on the fiber orientation angles. The external potential acting on the space variables is supposed to be one order of magnitude stronger than the one acting on the fiber rotations: $U^0 = O(1)$, $U^1 = O(\varepsilon)$. The strength of the alignment potential is supposed to be (large) of order $O(\varepsilon^{-1})$. The intensity of the alignment potential between linked fibers is supposed to be (small) of order $O(\varepsilon)$, and the diffusion coefficient and parameter ξ (ratio between the total number of fibers and total number of links) are supposed to stay of order 1. The main assumption in this scaling, which is introduced to simplify the analysis of the system, consists in supposing that the processes of linking and unlinking occur at a very fast time scale, i.e. $\nu_f, \nu_d = O(\frac{1}{\varepsilon^2})$.

It is noteworthy that these scaling hypothesis are done for technical reasons, and in this regime it has first been shown (see [9]) that the two particle distribution function g reduces to:

$$g^\varepsilon(x_1,\theta_1,\ell_1,x_2,\theta_2,\ell_2) = \frac{\nu_f}{\nu_d} f^\varepsilon(x_1,\theta_1) f^\varepsilon(x_2,\theta_2) \delta_{\bar{\ell}(x_1,\theta_1,x_2,\theta_2)}(\ell_1)\, \delta_{\bar{\ell}(x_2,\theta_2,x_1,\theta_1)}(\ell_2)$$
$$+ O(\varepsilon^2).$$

Note that as $\varepsilon \to 0$, the correlations built by the fiber links disappear since the two-particle distribution function then corresponds to the product of two one-particle distribution functions. This property comes from the fact that in this scaling limit, links appear and disappear almost instantaneously, making the timescale of the action of the restoring force much longer than the lifetime of a link.

Moreover, in this scaling limit, it can be shown that the one particle distribution function f^ε formally satisfies (neglecting the terms in $O(\varepsilon^2)$, see [9]):

$$\varepsilon \left[\partial_t f^\varepsilon - \partial_\theta \left(\left[\partial_\theta U^1 + \xi G[f^\varepsilon](x,\theta) \right] f^\varepsilon \right) - \mu d \Delta_x f^\varepsilon \right] = Q(f^\varepsilon), \tag{5.9}$$

where $Q(f)$ is the following collision operator:

$$Q(f) = d\partial_\theta^2 f + \xi \partial_\theta (\partial_\theta \Phi[f] f), \tag{5.10}$$

and

$$\Phi[f^\varepsilon](x_1, \theta_1) = C_1 \int_{-\frac{\pi}{2}}^{\frac{\pi}{2}} \sin^2(\theta - \theta_2) f^\varepsilon(x_1, \theta_2) \frac{d\theta_2}{\pi} \tag{5.11}$$

$$G[f^\varepsilon](x_1, \theta_1) = C_2 \sum_{i,j=1}^{2} \frac{\partial^2}{\partial x_i \partial x_j} \int_{-\frac{\pi}{2}}^{\frac{\pi}{2}} f^\varepsilon(x_1, \theta_2) B_{ij}(\theta_1, \theta_2) \frac{d\theta_2}{\pi}, \tag{5.12}$$

$$C_1 = \frac{\alpha L^2 v_f}{2 v_d}, \quad C_2 = \frac{\alpha L^4 v_f}{48 v_d}, \tag{5.13}$$

with α the alignment force intensity (see Eq. 5.4). Finally,

$$B(\theta_1, \theta_2) = \sin 2(\theta_1 - \theta_2)[\omega(\theta_1) \otimes \omega(\theta_1) + \omega(\theta_2) \otimes \omega(\theta_2)] = \left(B_{ij}(\theta_1, \theta_2)\right)_{i,j=1,2}, \tag{5.14}$$

where $A \otimes B$ is the tensor product between vectors A and B: $(A \otimes B)_{ij} = A_i B_j$. Equation (5.9) shows that the interactions at leading order are contained in the so-called collision operator $Q(f)$ (Eq. (5.10)), which expresses that the alignment potential (contained in the functional $\Phi[f](x, \theta)$) is counter-balanced by the diffusion term which tends to spread the particles isotropically on the sphere. It is noteworthy that the alignment force is local in space and consists of a sum of elementary alignment forces generated by intersecting fibers. The other terms (left-hand side of Eq (5.9)) act at lower order ε and contain the contribution of the external potential in orientation, the next order contribution of the alignment force between linked fibers (term $G[f]$) and the diffusion in space (due to fiber random motion in the microscopic model).

We now aim to study the limit $\varepsilon \to 0$ in Eq. (5.9). Formally, if we let $f^\varepsilon \to f$, as $Q(f^\varepsilon) = O(\varepsilon)$, we have $Q(f) = 0$. Therefore, f is an equilibrium of the collision operator, and consists of a Von Mises distribution $f(x, \theta, t)$ (see [9]):

$$f(x, \theta, t) = \rho(x, t) M_{\theta_0(x,t)}(\theta)$$

$$M_{\theta_0(x,t)}(\theta) = \frac{e^{r \cos 2(\theta - \theta_0(x,t))}}{\int_{-\frac{\pi}{2}}^{\frac{\pi}{2}} e^{r \cos 2\theta} \frac{d\theta}{\pi}},$$

where $\rho(x, t)$ is the fiber density, $\theta_0(x, t)$ the fiber local orientation and r the order parameter given by:

$$r = \frac{\xi \alpha L^2 \rho(x,t) c(r) v_f}{4 d v_d}, \quad c(r) = \int_{-\frac{\pi}{2}}^{\frac{\pi}{2}} \cos 2\theta M_0(\theta) \frac{d\theta}{\pi}.$$

Note that $c(r)$ does not depend on θ_0. Therefore in the limit $\varepsilon \to 0$, the one particle distribution function is fully described by the density $\rho(x, t)$ and the mean local orientation $\theta_0(x, t)$. Therefore, we need to find two equations to determine $\rho(x, t)$ and $\theta_0(x, t)$. Let us note that the local fiber number is the only quantity conserved by the interactions. In particular, there is no momentum conservation. Therefore, the only collision invariants of the collision operator are the constants. The integration of Eq. (5.9) against these invariants does not allow us to find the evolution equation for the mean orientation. In order to obtain an equation for θ_0, inspired from [8], the concept of Generalized Collision Invariants (GCI), i.e. of collision invariants when acting on a restricted subset of functions f, is used. Thanks to this new concept and in the case of a homogeneous fiber distribution $\rho(x, t) = \rho_0$ (see [9]), the local fiber orientation $\theta_0(x, t)$ solves:

$$\partial_t \theta_0 - \sum_{i,j=1}^{2} \partial_{x_i}\left(a_{ij}(\theta_0)\partial_{x_j}\theta_0\right) + \alpha_5 h(\theta_0) = 0, \tag{5.15}$$

where $a_{ij} \in \mathbb{R}$ for $i, j = 1, 2$ are the coefficients of a 2×2 matrix A such that:

$$A(\theta) = \begin{pmatrix} \alpha_2 - \alpha_3 \cos 2\theta & -\alpha_3 \sin 2\theta \\ -\alpha_3 \sin 2\theta & \alpha_2 + \alpha_3 \cos 2\theta \end{pmatrix}. \tag{5.16}$$

The coefficients $\alpha_1, \alpha_2, \alpha_3$ and α_5 are fully determined by the parameters r, d and L. Their expression is omitted here for the sake of simplicity (see [9]). Finally, the function h is the macroscopic counterpart of the external potential U and reads:

$$h(\theta_0) = \int_{-\frac{\pi}{2}}^{\frac{\pi}{2}} U'(\theta) M_{\theta_0}(\theta) \frac{d\theta}{\pi}. \tag{5.17}$$

In [20] it has been shown that, in the stationary case, Eq. (5.15) is a quasi-linear elliptic equation, and the existence of solutions was proven under structural conditions for the external potential $h(\theta_0)$.

Numerical simulations of the stationary solutions of Eq. (5.15) on a 2D square domain with Dirichlet boundary conditions have been performed in [20]. For an external rotation potential U forcing the fibers to reach orientation $\pm\frac{\pi}{2}$, the fiber network could be seen as a continuum medium subjected to compressive stress. In this case, a buckling phenomenon was observed, corresponding to an instability characterized by a sudden sideways failure of the structure subjected to high compression. The boundary conditions determined the mode of bending, the load corresponded to the external potential force and the point of failure depended on the elastic modulus of the fiber network contained in parameter α_3. In Fig. 5.2, we show an example of simulation obtained for the following external force:

$$U(\theta) = c_u \sin^2(\theta - \frac{\pi}{2}),$$

Fig. 5.2 Fiber orientation as function of x for four different values of the external potential c_u: $c_u = 0$ (black dots), $c_u = 0.5$ (blue curve), $c_u = 1$ (red curve) and $c_u = 10$ (green curve). Boundary conditions $\theta(-0.5, y) = 0.98\frac{\pi}{4}$ and $\theta(0.5, y) = -\frac{\pi}{4}$ for all $y \in [-0.5, 0.5]$

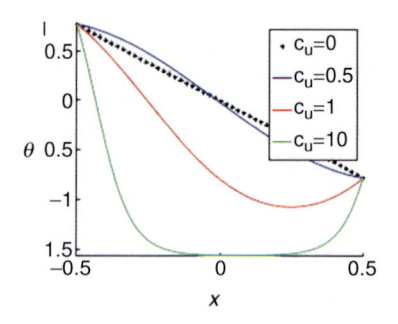

where c_u is the intensity of the force. We consider a 2D square domain $[-0.5, 0.5] \times [-0.5, 0.5]$ with Dirichlet boundary conditions on the left and right sides of the domain ($\theta = 0.98\frac{\pi}{4}$ and $\theta = -\frac{\pi}{4}$ respectively), and periodic boundary conditions on the top and bottom. Details on the numerical method can be found in [20]. In this setting, we can show that the solution θ does not depend on the y-direction and we plot in Fig. 5.2 the fiber orientation as function of x for $y = 0.2$, for four different values of the external potential intensity c_u.

As shown by Fig. 5.2, two stationary states are obtained when introducing the rotation potential: (1) a symmetric state with respect to the x-direction (black dots and blue curve), and (2) an asymmetric state in which all the fibers are oriented in $-\frac{\pi}{2}$ (red curve). There exists a critical c_u for which the solutions are in the unstable configuration: a slight increase of c_u leads the solution to buckle and change for configuration (2). These simulations highlight the physical relevance of the macroscopic model to describe interconnected networks as elastic materials with internal resistance depending on the density of the fiber links.

Moreover, numerical simulations showed a very good agreement between the macroscopic model and the microscopic formulation in a well chosen regime of fiber linking/unlinking. We illustrate this in Fig. 5.3, where we show simulations of the microscopic model given by Eqs. (5.1)–(5.2) rescaled with the scaling of Sect. 5.4 with $\varepsilon = \frac{1}{4}$, and small linking/unlinking ratio $\frac{v_f}{v_d} = 0.1$. Figure 5.3 (A1)–(A3) show simulations for increasing values of the noise d: $d = 10^{-4}$, $d = 10^{-3}$ and $d = 5.10^{-3}$, and external potential $c_u = 0.01$. In Fig. 5.3b we present the simulations for $c_u = 0.1$. For each, we show the microscopic simulation at equilibrium, and we plot the profiles of θ as function of x, averaged over the y-direction. Black curves correspond to the solutions of the microscopic model, red curves are the profiles of the solutions of the macroscopic one. For small d and small external potential $c_u = 0.01$ (A1,A2), fibers are oriented towards $\pm\frac{\pi}{2}$ on the left and right hand sides of the domain, with a zone of fibers horizontally oriented in the center, as predicted by the macroscopic model. For increasing d (A3), the fibers are disorganized and have mean orientation $\pm\frac{\pi}{2}$. The macroscopic model captures the same features for the same parameters.

As depicted in Fig. 5.3, we obtain a very good agreement between the microscopic and macroscopic simulations for a small linking/unlinking ratio.

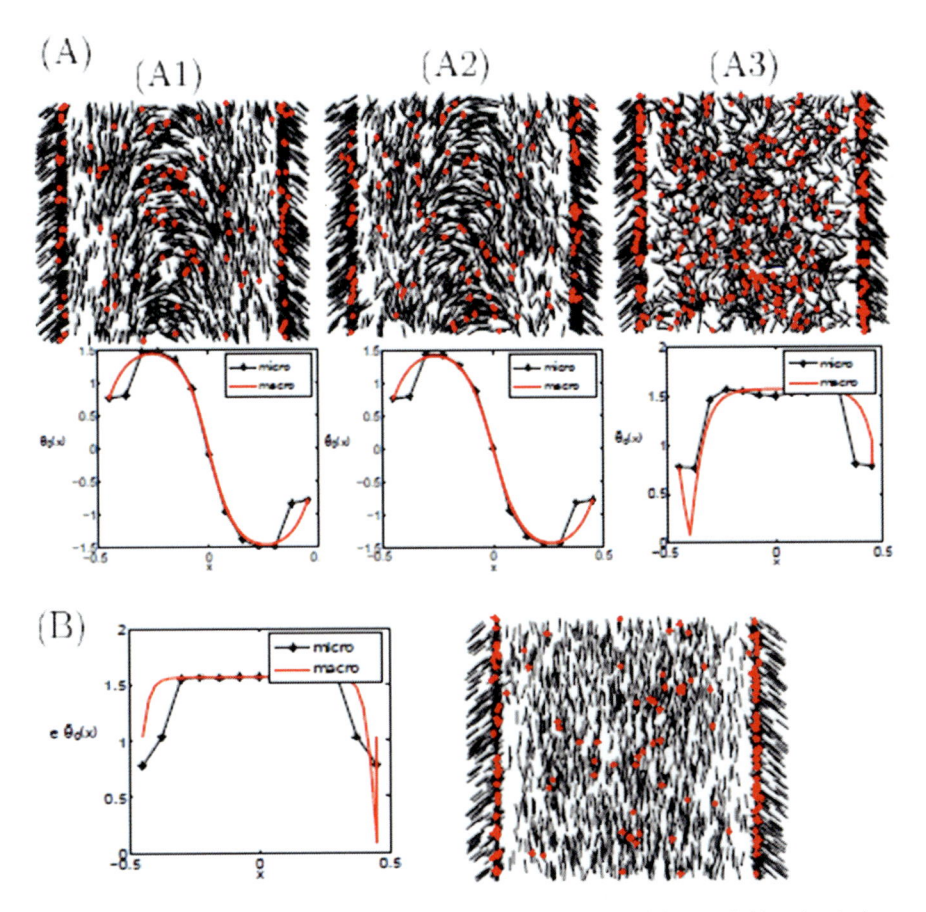

5 Modelling Tissue Self-Organization: From Micro to Macro Models

Fig. 5.3 Simulations of the microscopic model with external potential $c_u = 0.01$ and ratio $\xi = 0.1$. From A1 to A3: for increasing values of the noise d: $d = 10^{-4}$, $d = 10^{-3}$ and $d = 5.10^{-3}$. First line: simulation at equilibrium, second line: profiles of the solutions to the microscopic model averaged over the y-direction and over 10 simulations (black curves), and profiles of the solutions to the macroscopic model (red curves). For small d and small external potential $c_u = 0.01$ (A1,A2), fibers are oriented towards $\pm\frac{\pi}{2}$ on the left and right hand sides of the domain, with a zone of fibers horizontally oriented in the center, as predicted by the macroscopic model. For increasing d (A3), the fibers are disorganized and have mean orientation $\pm\frac{\pi}{2}$. The macroscopic model captures the same features for the same parameters. (B) Case $c_u = 0.1$ and $d = 10^{-3}$. In this case, all the fibers reach orientation $\pm\frac{\pi}{2}$ due to the large intensity of the external potential for both models. (**a**) Small external potential $c_u = 0.01$. (**b**) Large external potential $c_u = 0.1$

However, it was shown that the fiber links density has a strong impact on the final structures obtained by the microscopic model that the macroscopic model does not capture. This is due to the fact that the scaling supposes that the linking/unlinking process is quasi instantaneous. This assumption makes the action of the links vanish in the macroscopic model, and no memory effect of the fiber cross-links remains. Works are in progress to better account for these effects in the macroscopic model.

These simulations are a first step towards the validation of the macroscopic model for interconnected fibers and its derivation from a microscopic description. In conclusion, we give some exciting perspectives of this study.

5.5 Conclusion

The works presented in this chapter are, to our knowledge, the first attempt to derive as rigorously as possible a macroscopic model for temporarily cross-linked fibers interacting through alignment at their links from a microscopic model. The kinetic model obtained in the limit of a large number of individuals of the microscopic model involves two distribution functions: the fiber distribution function and the cross-link distribution function, and is closed at the level of the two-particle distribution function. The diffusive limit of the kinetic model in the regime of instantaneous fiber linking/unlinking leads to a system of two coupled nonlinear diffusion equations for the fiber density and mean orientation. In the homogeneous density case, physical properties of the solutions of the macroscopic model have been observed and the numerical comparison between the macroscopic model and the microscopic one has shown the relevance of the model in an appropriately chosen fiber linking/unlinking regime.

These works present numerous exciting perspectives in the mathematical and biological fields. Mathematically, rigorously proving the convergence of the particle model towards the mean-field limit or proving existence and uniqueness of smooth solutions for the macroscopic diffusion system with non-homogeneous fiber density are immediate perspectives. In the latter case, the much more complex system of two coupled highly non linear equations—for fiber density and local mean orientation—requires the development of advanced numerical methods. Further perspectives on this model include numerical simulations of the complete kinetic model and comparisons with the microscopic formulation. Finally, the establishment of a hydrodynamic scaling based on a more realistic assumption for fiber linking/unlinking dynamics would enable to understand how the presence of fiber links affects the macroscopic dynamics.

From the biological viewpoint and in the long-term, the hope is to couple the macroscopic model for fiber networks with a macroscopic model for the cells. The resulting coupled model will provide a complete "synthetic tissue" model, i.e. a large scale counterpart of the agent-based tissue model described in [21]. It will serve for the investigation of large scale effects in general tissue homeostasis. Obviously, biological relevance will require to extend the models to three spatial dimensions. As the present derivation could be extended to the 3D case without further complications, the main challenge lies in the modelling of the fibers and the links. For fibers, one could consider ellipsoids and further development is needed for the fiber links, which would lead to a different source term in the macroscopic model.

Data Statement No new data was collected in the course of this research.

Acknowledgements This work has been supported by the "Engineering and Physical Sciences Research Council" under grant ref: EP/M006883/1 and by the National Science Foundation under NSF Grant RNMS11-07444 (KI-Net). Pierre Degond acknowledges support from the Royal Society and the Wolfson foundation through a Royal Society Wolfson Research Merit Award. Pierre Degond is on leave from CNRS, Institut de Mathématiques de Toulouse, France. Diane Peurichard acknowledges support by the Vienna Science and Technology Fund (WWTF) under project number LS13 029.

References

1. Alonso R, Young J, Cheng Y (2014) A particle interaction model for the simulation of biological, cross-linked fibers inspired from flocking theory. Cell Mol Bioeng 7(1):58–72
2. Alt W, Dembo M (1999) Cytoplasm dynamics and cell motion: two phase flow models. Math Biosci 156:207–228
3. Astrom JA, Kumar PBS, Vattulaine I, Karttunen M (2005) Strain hardening in dense actin networks. Phys Rev E 71:050901
4. Baskaran A, Marchetti MC (2008) Hydrodynamics of self-propelled hard rods. Phys Rev E 77:011920
5. Bertin E, Droz M, Gregoire G (2009) Hydrodynamic equations for self-propelled particles: microscopic derivation, stability analysis. J Phys A Math Theor 42:445001
6. Camazine S, Deneubourg JL, Franks NR, Sneyd J, Theraulaz G, Bonabeau E (2001) Self-organization in biological systems. Princeton University Press, Princeton, NJ
7. Chaudury O, Parekh SH, Fletcher DA (2007) Reversible stress softening of actin networks. Nature 445:295–298
8. Degond P, Motsch S (2008) Continuum limit of self-driven particles with orientation interaction. Math Models Methods Appl Sci 18:1193–1215
9. Degond P, Delebecque F, Peurichard D (2016) Continuum model for linked fibers with alignment interactions. Math Models Methods Appl Sci 26:269–318
10. DiDonna BA, Levine A (2006) Filamin cross-linked semiflexible networks: fragility under strain. Phys Rev Lett 97(6):068104
11. Engwer C, Hillen T, Knappitsch M, Surulescu C (2015) Glioma follow white matter tracts: a multiscale DTI-based model. J Math Biol 71(3):551–82
12. Frouvelle A (2012) A continuum model for alignment of self-propelled particles with anisotropy and density-dependent parameters. Math Models Methods Appl Sci 22:1250011
13. Ginelli F, Peruani F, Bär M, Chaté H (2010) Large-scale collective properties of selfpropelled rods. Phys Rev Lett 104:184502
14. Ha SY, Tadmor E (2008) From particle to kinetic, hydrodynamic descriptions of flocking. Kinet Relat Models 1:415–435
15. Head DA, Levine AJ, MacKintosh FC (2003) Distinct regimes of elastic response, deformation modes of cross-linked cytoskeletal, semiflexible polymer networks. Phys Rev E 68:061907
16. Hillen T (2006) M5 mesoscopic and macroscopic models for mesenchymal motion. J Math Biol 53:585–616
17. Ilina O, Friedl P (2009) Mechanisms of collective cell migration at a glance. J Cell Sci 122:3203–3208
18. Joanny JF, Jülicher F, Kruse K, Prost J (2007) Hydrodynamic theory for multi-component active polar gels. New J Phys 9:422

19. Karsher H, Lammerding J, Huang H, Lee RT, Kamm RD, Kaazempur-Mofrad MR (2003) A three-dimensional viscoelastic model for cell deformation with experimental verification. Biophys J 85:3336–3349
20. Peurichard D (2016) Macroscopic model for linked fibers with alignment interactions: existence theory and numerical simulations. SIAM Multiscale Model Simul 14:1175–1210
21. Peurichard D et al (2017) Simple mechanical cues could explain adipose tissue morphology. J Theor Biol 429:61–81
22. Taber LA, Shi Y, Yang L, Bayly PV (2011) A poroelastic model for cell crawling including mechanical coupling between cytoskeletal contraction and actin polymerization. J Mech Mater Struct 6:569–589
23. Vicsek T, Zafeiris A (2012) Collective motion. Phys Rep 517:71–140

Chapter 6
A Multiscale Modeling Approach to Transport of Nano-Constructs in Biological Tissues

Davide Ambrosi, Pasquale Ciarletta, Elena Danesi, Carlo de Falco, Matteo Taffetani, and Paolo Zunino

6.1 Biophysics of Cancer

Transport phenomena play a fundamental role in the development of cancer. At different phases of cancer disease, tumors use mass transport to interact with the surrounding environment [30]; the propagation of growth signals or the invasion of the surrounding tissue by initiation of angiogenesis are essentially regulated by transport phenomena. Transport phenomena are also at the basis of the pharmacological treatment of cancer. The vascular network is a natural therapeutic option to target vascularized tumors. Nevertheless, the success of anticancer therapies in treating cancer cells *in vivo* is limited by their inability to reach their target in adequate quantities [31]. An agent that is delivered intravenously reaches cancer cells exploiting the vascular flow, then it crosses the vessel wall and diffuses through the tissue interstitium. Each of these steps can be seen as a barrier to delivery. In addition, even delivered molecules may bind to constituents of the extracellular matrix and be metabolised by cells.

D. Ambrosi · P. Ciarletta · E. Danesi · C. de Falco · P. Zunino (✉)
MOX, Dipartimento di Matematica, Politecnico di Milano, P.zza L. Da Vinci 32, 20133 Milano, Italy
e-mail: davide.ambrosi@polimi.it; pasquale.ciarletta@polimi.it; elena.danesi@polimi.it; carlo.defalco@polimi.it; paolo.zunino@polimi.it

M. Taffetani
Mathematical Institute, University of Oxford, Andrew Wiles Building, Radcliffe Observatory Quarter, Woodstock Road, Oxford OX2 6GG, UK
e-mail: matteo.taffetani@maths.ox.ac.uk

© Springer International Publishing AG, part of Springer Nature 2017
A. Gerisch et al. (eds.), *Multiscale Models in Mechano and Tumor Biology*, Lecture Notes in Computational Science and Engineering 122, https://doi.org/10.1007/978-3-319-73371-5_6

The characteristic traits of cancer can be seen as the emergent behavior of a cascade of phenomena that propagate from the molecular scale, through the cell and the tissue microenvironment, up to the systemic level. Transport phenomena in the capillary network (the *microenvironment* or *microscale*) play a key role in this sequence of effects. In particular, the alterations of the capillary phenotype of a tumor significantly affect the drug delivery process [7]: blood vessels in tumors are leakier and more tortuous than the normal microvasculature and the pressure generated by the proliferating cells reduces tumor blood and lymphatic flow. These alterations lead to an impaired blood supply and abnormal tumor microenvironment, characterized by hypoxia and elevated interstitial fluid pressure. The reduced pressure gap across the vessel walls is sometimes even reversed in sign and greatly reduces the ability to deliver drugs.

The mathematical modelling of interstitial flow can be an important support to tackle these issues. The objective of this work is to illustrate a multiscale mathematical model to investigate the diffusion and transport properties of nanoparticles after nanofluidic injection into a living tissue. Our multiscale approach is based on two coupled models. The macroscopic model considers the tissue as a porous medium, and accounts for convection, diffusion and absorption of the particles, by means of a continuous spatial concentration. The microscopic model studies the motion of the particles in the extra-cellular space and their interaction with the targeted cell surface. We follow the definition of characteristic scales proposed in [53]:

(i) The tissue scale (cm-tens of s), where the structured heterogeneities within the living material can be separately recognized;
(ii) The extravascular space (mm-s), where the matter transfers from blood in the capillary network to the extracellular matter and the cells;
(iii) The cellular level (μm-ms), in which nanoparticles/nanovectors diffuse and are advected by the flow or up-taken by the cells, behaving like a suspension within a fluid domain;
(iv) The sub-cellular level (up to hundreds of nm-μs), where the biochemical interactions, such as the particles uptaking or cells regulation, become the dominant processes.

The models labelled as *macroscale* span the scales between (i) and (ii), while the *microscale* typically involves phenomena of (iii) and (iv).

6.1.1 An Overview of Transport Phenomena in Tumors

Several mathematical models have been proposed in recent years to describe transport phenomena in tumor tissues, and different types of particles have been used in cancer treatment. In most cases, the tumor is assumed to be spherical, in analogy with the multi-cellular tumor spheroid model, an assumption that allows to exploit the radial symmetry; in general the significant parameters are considered constant and derived experimentally.

Banerjee et al. [3] described the penetration of monoclonal antibodies in a tumor nodule surrounded by healthy tissue. The nodule is assumed to be spherical and embedded in a tissue described by a macroscale model: the concentration of free antibodies in the external extracellular space is determined by the interaction between diffusion, uptake from blood and efflux to lymphatics. The distance between blood and lymphatic microvessels is assumed to be small in comparison with the tumor nodule and then uptake and efflux can be described by uniform rate constants. Exploiting the radial symmetry, the concentration of antibodies A_b in the healthy tissue is described as

$$\frac{\partial A_b}{\partial t} = D_e \frac{1}{r^2} \frac{\partial}{\partial r} \left(r^2 \frac{\partial A_b}{\partial r} \right) + \kappa A_{b,p} - \Lambda A_b, \qquad (6.1)$$

where $A_{b,p}$ is the concentration of antibodies in the blood, r is the radial coordinate, t is time, D_e is the diffusion coefficient in the healthy tissue, κ is the constant rate of transcapillary exchange, Λ is the constant rate of lymphatic flux from normal interstitium.

In the pre-vascular stage of the cancer the presence of lymphatic and blood vessels is neglected and can be described as in van Osdol [56] and Graff and Wittrup [29]. In the former, the concentration of free antibodies is coupled to the concentration of free antigens and to the concentration of the antigen-antibody complex B by the reaction terms,

$$\frac{\partial A_b}{\partial t} = D \frac{1}{r^2} \frac{\partial}{\partial r} \left(r^2 \frac{\partial A_b}{\partial r} \right) - k_f A_b A_g + k_r B, \qquad (6.2a)$$

$$\frac{\partial A_g}{\partial t} = n(-k_f A_b A_g + k_r B), \qquad (6.2b)$$

$$\frac{\partial B}{\partial t} = k_f A_b A_g - k_f B, \qquad (6.2c)$$

$$A_g + n B = A_{g,0}, \qquad (6.2d)$$

where k_f and k_r are constants that describe the association and dissociation rates between antibodies and antigens, $n = 2$ is the binding valence of antibodies, and $A_{g,0}$ is the total concentration of antibodies. The authors considered symmetric Dirichlet boundary conditions and a condition of interface between the tumor and the normal tissues. All constants in the model were determined experimentally; among all the parameters, the diffusion coefficient is very difficult to measure, thus explaining the large variability in the values that can be found in the literature. Numerical simulations show that the values chosen for the diffusion coefficient and the density of the binding sites in the tumor (in this case the antigens) play a fundamental role in the diffusion of the particles.

Graff and Wittrup [29] used a very similar model to describe the diffusion of antibodies in tumor spheroids. Their results are compared with those experimentally obtained in tumor spheroids and no healthy tissue is considered. The study focuses on the optimization of the affinity antibody/antigen. In previous simulations [56], it has been shown that the high affinity antibody/antigen is responsible for the formation of a *binding site barrier*: the front of free antibodies advances inside the spheroid until their concentration in plasma falls below a certain level. If the front stops when antibodies have not yet reached the center of the spheroid, they create a "barrier". The study shows that antibodies with lower affinity have a greater penetration ability, but their therapeutic efficacy decreases. Also in this model the association and dissociation rates are assumed to be constant and the convection within the tumor is neglected.

Ward et al. [58] and Norris et al. [40] propose a refined model for the diffusion of chemotherapic drugs in tumor spheroids and used it to compare the effectiveness of various techniques of drug release. The numerical analysis applies to a system of four one-dimensional equations in radial coordinates where different fields are considered: the density of living cells, the concentration of nutrients, the concentration of drug and the convective flow of the cells in the spheroid. The reaction terms account for mitosis and apoptosis rates of the cells: the former according to a generalized Michaelis-Menten kinetics, while the rate of cell death depends on the concentration of drug.

The models presented above are designed for the diffusion and transport of macromolecules and antibodies, and they can not be directly applied to the diffusion and transport of nanoparticles: nanoscale objects are subject to strong interaction with the cellular surface, that may cause further absorption with respect to the simple molecular degradation. Moreover the diffusivity of the nanoparticles is smaller than for macromolecules, since the Brownian motion, due to the collisions of the particles with the molecules of the fluid, is inversely proportional to the particles size. Goodman et al. [28] analyzed the diffusion of nanoparticles for size between 20 and 200 nm and developed a mathematical model which takes into account the inhomogeneities inside the tumor spheroids. They assume that the porosity ϵ depends on the radial coordinate and such a generalization allows the application of the model to spheroids treated with collagenase. In fact, it is experimentally known that the treatment of spheroids with ECM-degrading enzymes, such as collagenase and ialuronidase, increases of the pore volume (as shown in Fig. 6.1) and therefore facilitates the diffusion of the particles.

The model by Goodman et al. describes the evolution of the molar concentration C of free particles, the concentration of bound particles C_b, the concentration of available binding sites on the cell surface C_{bs}, the concentration of internalized particles C_i. The corresponding balance equations are

$$\frac{\partial C}{\partial t} = \frac{1}{r^2}\frac{\partial}{\partial r}\left[D\epsilon r^2 \frac{\partial}{\partial r}\left(\frac{C}{\epsilon}\right)\right] - k_a C_{bs}\frac{C}{\epsilon} + k_d C_b, \qquad (6.3a)$$

$$\frac{\partial C_b}{\partial t} = k_a C_{bs}\frac{C}{\epsilon} - k_d C_b - k_i C_b, \qquad (6.3b)$$

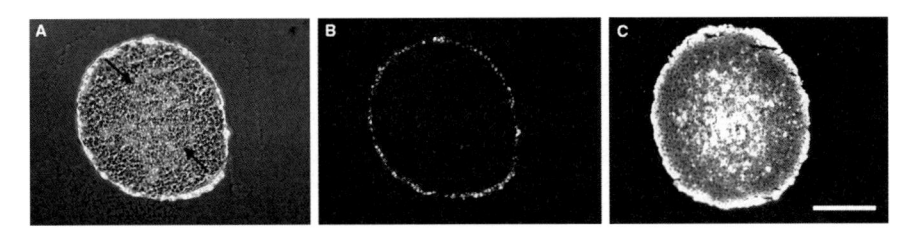

6 A Multiscale Modeling Approach to Transport of Nano-Constructs in...

Fig. 6.1 Sections of a spheroid exposed to 40 nm fluorescent nanoparticles. (**a**) Section shown as a phase contrast; arrows indicate the approximate boundary of the necrotic core. (**b**) Section shown as a fluorescent image. (**c**) Fluorescent image of a spheroid coincubated with 0.076 mg/mL collagenase and nanoparticles. Scale bar is 200 μm [28]

$$\frac{\partial C_{bs}}{\partial t} = -k_a C_{bs} \frac{C}{\epsilon} + k_d C_b + k_i C_b, \tag{6.3c}$$

$$\frac{\partial C_i}{\partial t} = k_i C_b, \tag{6.3d}$$

where k_a is the rate of association, k_d the rate of dissociation, k_i the rate of internalization. We point out that, in this setting, the porosity of the material ϵ is not an asymptotically small parameter and we recall that it may depend on the radial coordinate, namely $\epsilon = \epsilon(r)$. The initial and boundary conditions are given by

$$\begin{aligned} C(0, r) = C_b(0, r) = C_i(0, r) = 0, &\quad 0 \leq r < R, \\ C_{bs}(0, r) = C_{bs,0}, &\quad 0 \leq r < R, \\ C(t, R) = C_0 \epsilon(R), &\quad t > 0, \\ \frac{\partial}{\partial r}\left(\frac{C}{\epsilon}\right)(t, 0) = 0, &\quad t > 0, \end{aligned} \tag{6.4}$$

where R is the radius of the spheroid, C_0 is the concentration of particles outside the spheroid and $C_{bs,0}$ is the initial concentration of the binding sites inside the spheroid.

The model is formally analogous to the one by Graff and Wittrup [29], but the particles internalization is also considered. The article highlights the difficulty in finding an expression for the diffusion coefficient in tumor tissue that should synthetically represent various factors (such as the arrangement of the cells and the composition of the extracellular matrix). A possible expression for the diffusion coefficient in porous media is given by Goodman et al. [28]

$$D = D_0 \frac{L(\lambda)}{F \tau(\epsilon)}, \tag{6.5}$$

where

$$D_0 = \frac{K_B T}{6\pi \eta r_p} \quad \text{and} \quad \lambda = \frac{r_p}{r_{pore}}. \tag{6.6}$$

Here D_0 is the diffusion coefficient in an unbounded liquid medium, r_p is the particle radius, r_{pore} is the effective pore size, η is the fluid viscosity, K_B is the Boltzmann constant, T is the absolute temperature, $L(\lambda)$ is the factor responsible for hydrodynamic and steric reduction of the diffusion coefficient in the pore, $\tau(\epsilon)$ is the tortuosity and $F > 1$ is a shape factor that accounts for the hindrance in the pores. While this model considers non-specific interactions, it shows that the diameter of the nanoparticles is a fundamental parameter for the diffusion of the drugs and introduces the space-dependent porosity ϵ as an approach for modeling the heterogeneity of the tissue (Fig. 6.2).

The model proposed by Goodman has been seminal in the field. Different expressions for the diffusion coefficient have been later proposed by Florence [23] and Gao et al. [25]. The former author proposes various expressions for the diffusion coefficient, taking into account the obstacles present in the extracellular matrix. The second one compares the model-simulated results on the diffusion of

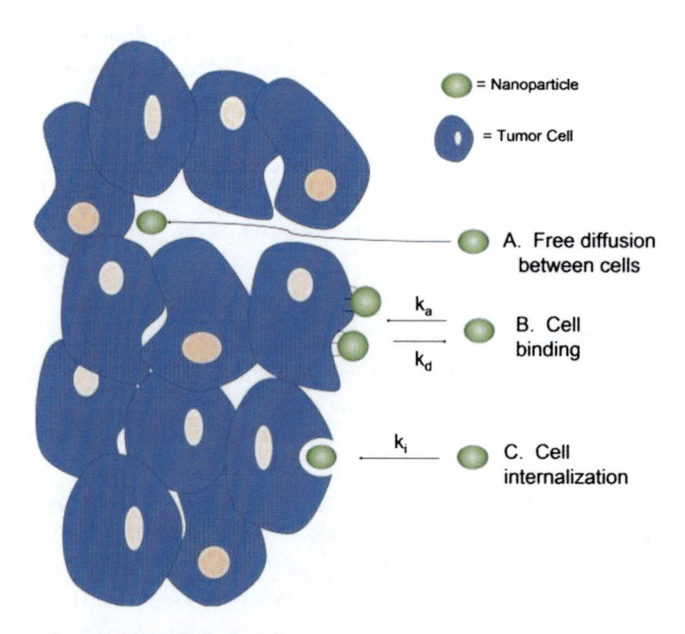

Fig. 6.2 Transport mechanisms governing nanoparticles penetration through solid tumors. Nanoparticles are transported through tumors by (A) free diffusion in extracellular space, and can be inhibited by (B) cell binding and/or by (C) cell internalization. The structure of nanoparticles can be tuned to alter their interactions with cells and the tumor bed, thus optimizing their transport through solid tumors [57]

nanoparticles with experimental results obtained in the case of three different types of nanoparticles with positive, negative and neutral charges, respectively.

Waite and Roth [57] applied a model similar to the one by Goodman to a particular kind of nanoparticles (dendrimers PAPAM) simulating the effect of the change in the number of targeting ligands (RGD peptides) on their surface. The affinity of these ligands with a malignant glioma cells had been demonstrated experimentally. This type of particles is positively charged and exerts electrostatic interactions with the cell membrane. The previous models are expanded to include two possible modes for the formation of bonds between nanoparticles and cells: through receptor-mediated interactions (between RGD targeting peptides and cell-surface integrin proteins) or non-specific ones (through electrostatic interactions). The differential system (6.3) is supplemented by the equations for the concentration of particles linked to specific binding sites and for the concentration of available specific binding sites. Graff and Wittrup have shown that low affinity antibodies have a more homogeneous distribution through solid tumours; in this paper a similar behavior is observed in the case of tumor-targeted nanoscale materials. Unlike in Goodman analysis, the porosity is assumed to be constant through the spheroid, as well as the diffusion coefficient and the association and dissociation rates, which are estimated through the comparison with experimental data.

A similar model was obtained in Ying et al. [59] for the transport of drugs through vesicles, which have been studied along several years for their ability to deliver drugs. In particular, various methods were evaluated for the calculation of the coefficient of association between receptors and specific ligands, considering also the possibility to use different types of ligands on a single vesicle. In this work, no formation of non-specific binding is considered.

At the sub-cellular level, the theory proposed by Decuzzi and Ferrari [17, 18] studies the specific interactions between cells and particles to assess the relation between the dynamics of absorption and the shape of the particles. The absorption rate of the particles is then a function of the binding energy, the density and mobility of receptors on the cell membrane and the shape of the particles.

Chou et al. [12] adopted a different point of view and constructed a small-scale model to study the evolution of macromolecule carriers in localized tumor tissues. While the articles presented above considered spheroids tumor or tumors in pre-vascular phase, in this case a convective flux of the interstitial fluid is taken into account.

Su et al. [52] studied a multiscale model for the transport and deposition of magnetite nanoparticles during the injection of nanofluids into tumor tissues. They track the trajectory of a particle in the extracellular matrix through the study of its dynamics and the interactions between the nanoparticle and a cell; this study allows to obtain the rate of particles deposition on a single cell, from which the volumetric absorption coefficient in the tissue can be computed and then used in a macroscale model.

6.2 A Microscale Approach to Transport of Nano-Constructs

6.2.1 Microscopic Model

In this section we restrict our analysis to the motion of nanoparticles in the extracellular space, around either a single cell or in a pore, with the aim of determining the single cell efficiency η_s as a function of the relevant biophysical phenomena. The hydrodynamic interactions between the particles can be neglected since the nanofluid is dilute and the nanoparticles behave as point-like, chemically inert, solid spheres.

Due to the small length-scales under considerations, the flow is characterized by a very small Reynolds number, it is laminar and obeys the Stokes equations

$$\begin{cases} \nabla p = \eta \Delta \mathbf{v}_f, \\ \nabla \cdot \mathbf{v}_f = 0, \end{cases} \tag{6.7}$$

where p is the pressure and $\mathbf{v}_f$ is the fluid velocity. In the following, we solve these equations for given pressure drop at the boundary, no slip conditions on the cell surface and boundary conditions dictated by the symmetry of the problem elsewhere.

The nanoparticles around a cell are subject to the drift of the flowing liquid, while their motion is affected by Brownian motion and other forces (see Sect. 6.2.1.2). Accordingly, the distribution of the nanoparticles can be modeled through either a discrete stochastic approach, which calculates the trajectory of the particles, or through its continuous limit, which studies the evolution of nanoparticle density in time and space.

Using the discrete viewpoint, the trajectory of a nanoparticle is described by the stochastic Langevin equation:

$$d\mathbf{r}_j = \left(\frac{D}{K_B T} \mathbf{F}^e + \mathbf{v} \right) \Delta t + (\Delta \mathbf{r})_j^B, \tag{6.8}$$

where $d\mathbf{r}_j$ is the displacement vector of the jth particle, D is the particle diffusivity tensor, $\mathbf{F}^e$ represents the sum of all the forces acting on the particle, $\mathbf{v}$ is the particle velocity, Δt is the time step and $(\Delta \mathbf{r})_j^B$ is the random Brownian displacement due to collisions between the particles and the fluid molecules surrounding them. Moreover K_B is the Boltzmann constant and T is the absolute temperature. Since nano-sized-particles have a small relaxation time, we can neglect inertia and assume that the particles relax to the fluid velocity almost instantaneously. Thus, the particle velocity $\mathbf{v}$ coincides with the fluid velocity.

Conversely, using a continuous viewpoint, the evolution of the nanoparticles concentration c can be described by the following advection-diffusion equation

$$\frac{\partial c}{\partial t} - \nabla \cdot \left(D \nabla c - \mathbf{v} c - \frac{1}{K_B T} D \mathbf{F}^e c \right) = 0, \tag{6.9}$$

where D is a diffusivity tensor. The next section is devoted to the identification of the major factors influencing the microscopic motion of the nanoparticles in the extracellular space: the hydrodynamic interactions and the forces acting on the nanoparticles.

6.2.1.1 Hydrodynamic Retardations

The diffusivity tensor D and the nanoparticle velocity **v** appear in both Eqs. (6.8) and (6.9). Since inertia is negligible, the nanoparticle velocity **v** can be assumed to coincide with the fluid velocity. Far from the cell surface, the perturbation of the nanoparticle on the flowfield can be neglected and the velocity of the fluid obeys to Stokes equation (6.7) without any feedback. Similarly, the nanoparticle diffusivity can be assumed to be isotropic according to Stokes-Einstein relation far away from the cell. Conversely, near the cell the hydrodynamic interactions between the particles and the cell surface significantly affect the motion of the fluid, thus influencing the induced particle velocity and diffusivity [26, 27]. Since the nanoparticles are much smaller than a cell and the perturbation to the fluid flow is significant only near the cell surface, this one can be approximated as a plane wall and the following relations can be introduced [19]

$$v_\| = f_3(\delta) v_{f,\|} \quad , \quad v_\perp = f_1(\delta) f_2(\delta) v_{f,\perp},$$
$$D_\| = f_4(\delta) D_0 \quad , \quad D_\perp = f_1(\delta) D_0, \tag{6.10}$$

where the subscripts $\|$ and $\perp$ indicate the components in the direction parallel and perpendicular to the surface, respectively. The scalar functions f_i, $i = 1, \ldots, 4$ depend on the normalized distance δ, defined as the ratio between the distance H of the particle center from the cell surface and the particle radius r_p. Their values can be fixed by estimating the hydrodynamic forces and torques acting on a particle near the surface and exploiting the equilibrium equations. In our simulations, we used the values obtained computing these functions by numerically solving four test cases and validating our results with the approximated expressions presented in [52].

6.2.1.2 Driving Forces

When a particle moves immersed in the fluid filling the extracellular space, several forces should be considered: drag, Basset force, Magnus force, Saffman force and buoyancy force. Nevertheless, all the mentioned forces can be neglected in our study due to the small size of the particles considered. In our microscopic model, we take into account only two type of interactions between the nanoparticles and the cell:

$$\mathbf{F} = \mathbf{F}_{lift} + \mathbf{F}_{DLVO}. \tag{6.11}$$

where $\mathbf{F}_{DLVO}$ represents the sum of Derjaguin, Landau, Verwey, and Overbeek (DLVO) forces and $\mathbf{F}_{lift}$ the lift force. All of these contributions act normally to the cell surface and are significant when the particle is close to the cell surface.

The DLVO theory quantitatively explains the aggregation of aqueous dispersions and is based on the combined effect of two forces: the van der Waals force and the electrostatic double layer force. The latter arises only between charged bodies through an electrolyte solution and it will be neglected in the following since we deal with non charged nanoparticles. The van der Waals intermolecular force is due to the interactions between the dipoles constituting the bodies and is given by:

$$\mathbf{F}_{DLVO} = -\frac{\partial U_{vdW}}{\partial h}, \tag{6.12}$$

where h is the distance between the two surfaces and U_{vdW} is the van der Waals interaction energy. An approximate expression for U_{vdW} between a sphere of radius r_p and a plane wall is given by Israelachvili [33]

$$U_{vdW} = -\frac{A_H}{6}\left(\frac{r_p}{h} + \frac{r_p}{2r_p + h} + \ln\frac{h}{2r_p + h}\right), \tag{6.13}$$

where A_H is the Hamaker constant.

The lift force acts on a nanoparticle which moves in a fluid flowing close to a surface: in this condition, the gradient of velocity tends to move the particle in the direction normal to the streamlines of the fluid flow. The lift force acting on a neutrally buoyant particle in a fluid having a linear shear flow profile near a wall is given by Cherukat and McLaughlin [11]

$$\mathbf{F}_{lift} = r_p \eta V Re I \mathbf{i}_\perp, \tag{6.14}$$

where I is such that

$$I = \left(1.7631 + 0.3561\zeta - 1.1837\zeta^2 + 0.845163\zeta^3\right) + \\ - \left(3.24139\zeta^{-1} + 2.6760 + 0.8248\zeta - 0.4616\zeta^2\right)\Lambda_G + \\ + \left(1.8081 + 0.879585\zeta - 1.9009\zeta^2 + 0.98149\zeta^3\right)\Lambda_G^2, \tag{6.15}$$

with

$$\zeta = \frac{r_p}{H}, \qquad \Lambda_G = \frac{Gr_p}{V}. \tag{6.16}$$

In Eqs. (6.14)–(6.16) V is a characteristic velocity, Re is the Reynolds number calculated using the particle radius, H is the distance of the particle center from the cell surface and G is the velocity gradient of the linear shear flow field in the direction normal to the surface.

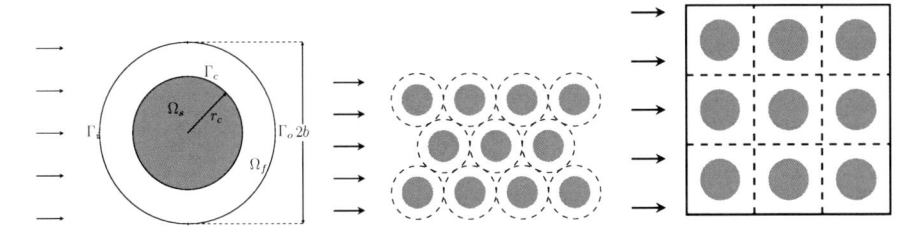

Fig. 6.3 Cell and fluid envelope in Happel's sphere-in-cell model (left) and corresponding hexagonal package (middle), highlighting the presence of voids in the microstructural domain. Cubic packing of cells in the case of spherical cells (right)

6.2.1.3 Microscopic Geometry

In the study of particles deposition onto solid surfaces, Happel's sphere-in-cell model has been extensively applied as a representation of the unit cell constituting the porous medium [39, 46, 55]. In this model the medium is represented as an assembly of identical spherical cells of radius r_c, each one surrounded by a shell of fluid (Fig. 6.3); the outer boundary is a concentric sphere of radius b, where b is chosen so that the overall porosity of the medium is preserved for the single cell, that is

$$b = r_c(1-\epsilon)^{-\frac{1}{3}}, \qquad (6.17)$$

as shown in Fig. 6.3(left). The axial symmetry of the problem allows an analytical solution of Stokes equation (6.7). In particular, the components of the fluid velocity can be obtained in spherical coordinates [19] as a function of the uniform fluid velocity U_∞ far away from the cell. Nonetheless, this model is not well representative of the real biological microstructure, since there is no package of single spherical Happel's elements which allows to fill the whole microstructured domain, as sketched in Fig. 6.3(middle). Consequently, the porosity of the medium is not the same as the one of the single microscopic domain.

Our modelling approach considers a cubic packing, as sketched Fig. 6.3(right), for which the definition of porosity is kept consistent between the microscopic and the macroscopic level. Since an analytic solution of Stokes equation is not available here, we first numerically solved the problem for the elementary cubic cell. Secondly, we considered two different microscopic geometries, mimicking the particles' absorption onto a cell surface and the extravasation over a capillary lumen. In the former case, we considered a spherical cell (Fig. 6.4(left)) and an ellipsoidal cell (Fig. 6.4(middle)) with cubic packing. In the latter, we investigated the motion of the nanofluid into a pore, considered both as a sinusoidal channel and a sinusoidal layer, whose two-dimensional projection in the x-y plane is shown in Fig. 6.4(right). For the sake of clarity, the sinusoidal pore is assumed symmetric for rotations around the x axis, while the sinusoidal layer has a translational symmetry

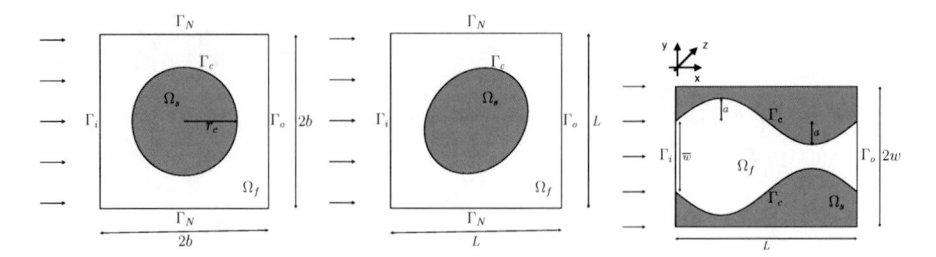

Fig. 6.4 Two-dimensional projection in the x-y plane of the unit spherical cell with cubic packing (left) and two-dimensional projection in the x-y plane of the unit ellipsoidal cell with cubic packing (middle). On the right we show the two-dimensional projection in the x-y plane of the sinusoidal pore (obtained by rotation invariance around the x axis) and the sinusoidal layer (translational symmetry with respect to the z axis)

Table 6.1 Main biological parameters used in the numerical simulations

Parameter	Description	Value
A_H	Hamaker constant	4×10^{-20} J
T	Temperature	310.15 K
d_p	Particle diameter	0.1–200 nm
ρ_f	Fluid density	997 kg m^{-3}
ρ_p	Particle density	1060 kg m^{-3}
η	Dynamic fluid viscosity	0.001 kg m^{-1} s^{-1}

with respect to the z axis. Finally, all the biological parameters implemented in the numerical simulations have been collected in Table 6.1.

6.2.2 Upscaling Method

Assuming that the porous tissue is homogeneous and composed by a periodic array of identical unit cells, we can relate the macroscopic model to the microscopic one, and in particular the local deposition rate coefficient k_f to the single cell efficiency η_s. In analogy with [45] we adopt the following relation between k_f and η_s

$$k_f = -\frac{u_x}{L_d} \ln(1 - \eta_s), \qquad (6.18)$$

where L_d is the length of the microscopic domain and u_x is the fluid velocity in the principal direction of flow (assumed to be constant, obtained for example by averaging the flow field over the unit cell).

6.2.3 Numerical Methods

For the solution of the Langevin equation (6.8) we used a Lattice Kinetic Monte Carlo method with a fixed uniform Cartesian grid. This approach exploits the high dilute assumption for the nanofluid and allows very efficient parallel simulation of particle trajectories due to the possibility to precompute all transition rates between lattice sites. The crucial aspect of the algorithm design is a suitable selection of transition times [36] that allows to avoid undesirable effects that are well known in the literature [20]. The continuum model (6.9) has been numerically solved by using a Finite Volume Method which combines Exponential Fitting for stabilization, in case of dominant advection, with the technique of [44], which allows to construct 3D control volumes that enforce a Discrete Maximum Principle. The velocity field appearing both in the discrete stochastic model of Eq. (6.8) and in the continuum model (6.9) has been calculated by a Finite Element solver for Stokes' equation based on the MINI algorithm of [2]. The implementation of the LKMC and FVM algorithms was performed based on in-house C++ libraries https://redmine.mate.polimi.it/projects/bim, (2016), while the Stokes solver was implemented in python using the DOLFIN library [35].

6.2.4 Numerical Results

The numerical simulation of our multiscale model investigate the effect of nanoparticles and tissue properties on the single cell absorption efficiency at the microscopic level and the consistency of the upscaling method at the macroscopic scale.

6.2.4.1 Sensitivity Analysis of Microscopic Effects

The many factors that affect the motion of the nanoparticles in the extracellular space at the microscopic level can be separately considered and evaluated in a numerical simulation. In Fig. 6.5, we show the effect of the hydrodynamic retardations on the single cell absorption efficiency: the results obtained for the Happel's sphere-in-cell model and the sinusoidal pore are compared. In both cases, the retardation reduces the single cell efficiency for all particles sizes, with a more significant difference arising for bigger nanoparticles.

In Fig. 6.6, the influence of each of the microscopic forces acting on the nanoparticles is reported. We can see that the effect of the lift force is almost negligible in the low Reynolds number regime of interest to our application, so that the most relevant contribution is given by the van der Waals force. Since the van der Waals force is always attractive, we calculate a growth of the absorption efficiency, which increases with the nanoparticle size.

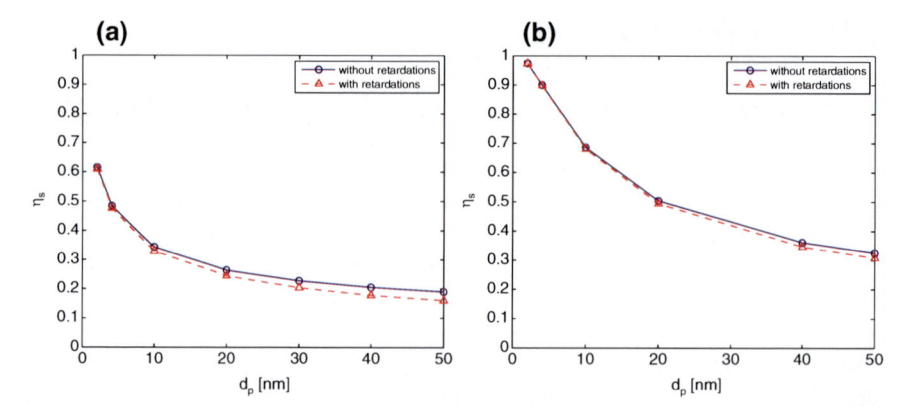

Fig. 6.5 Effect of hydrodynamic retardations on the single cell absorption efficiency, in the case of (**a**) Happel's sphere-in-cell model, (**b**) sinusoidal pore with $\tau = 0.2$

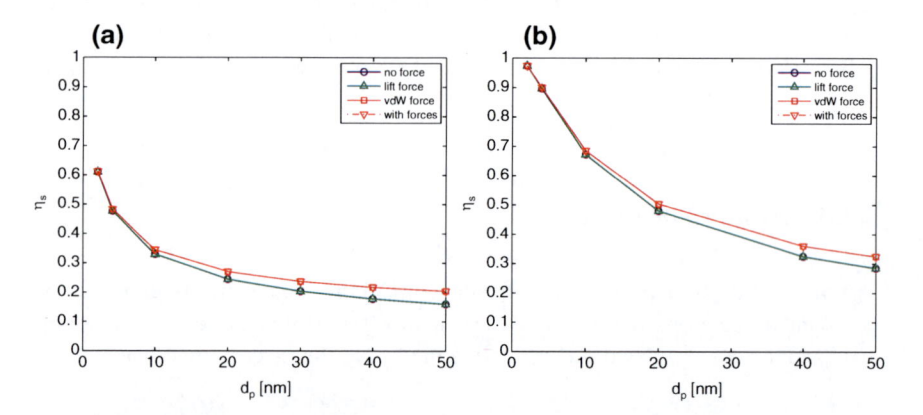

Fig. 6.6 Effect of van der Waals and lift forces on the single cell absorption efficiency, in the case of (**a**) Happel's sphere-in-cell model, (**b**) sinusoidal pore with $\tau = 0.2$

6.2.4.2 Nanofluid Motion Around a Cell

The motion of the nanofluid in the first system model influences the particle absorption onto a cell surface. We have compared the results for the Happel's sphere-in-cell model, obtained through simulations with the Monte Carlo method and the Finite Volume Method, with RT [46] and TE [55] semi-analytical correlations, versus the results presented in [52]. In Fig. 6.7, we report the absorption efficiency versus the nanoparticle diameter, showing that the cell efficiency decreases with increasing particle size. As expected, the effect of diffusion becomes smaller and the nanoparticles deviate less from the streamlines of the fluid flow. We remark that our numerical simulations yield a higher cell efficiency with respect to the literature values, except for the smaller particles.

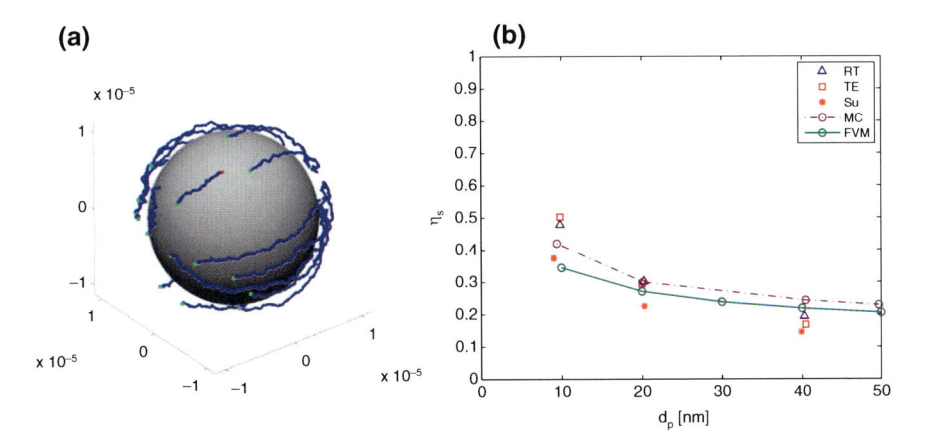

Fig. 6.7 Nanoparticles motion in the fluid layer in Happel's sphere-in-cell model. (**a**) Nanoparticles trajectories obtained through Monte Carlo simulations; (**b**) trend of single cell absorption efficiency versus particle size in the case of Happel's sphere-in-cell model. Here RT and TE represent the results of semi-analytical correlations, Su denotes the results in [52], MC and FVM denote the results obtained through our numerical simulations with Monte Carlo method and Finite Volume Method. $r_c = 10\,\mu\text{m}$, $\epsilon = 0.36$, $U_\infty = 3.7 \cdot 10^{-4}\,\text{m s}^{-1}$

Secondly, we have investigated the effect of the microscopic geometry on the single cell absorption efficiency, taking into account the more realistic cubic packing of the cubic unit cell. Since the efficacy of the FVM model has been validated against the MC one, we only consider the continuum approach. Then, we have first numerically computed the fluid velocity in the representative cubic domain solving Stokes equation (6.7) and then we have solved Eq. (6.9) for the nanoparticle concentration by applying a Finite Volume Method. We have compared the single cell efficiency obtained through the Happel sphere-in-cell model against the case of a spherical cell with cubic packing with r_c and b fixed and different values of porosity. Let us highlight that the minimum allowed porosity in the case of spherical cells with cubic packing is larger than in the case of Happel's model. Figure 6.8 shows that η_s is smaller for a sphere with cubic packing than for the Happel's model.

Thirdly, we have compared the efficiency of a single spherical cell with an ellipsoidal cell with cubic packing at fixed porosity $\epsilon = 0.6649$. The resulting curves of single cell absorption efficiency as a function of the nanoparticle size are plotted in Fig. 6.9. In both cases the cell efficiency decreases for increasing particle size, being almost the same for the smallest nanoparticles. For the spherical cell, the efficiency also increases for increasing nanoparticle size.

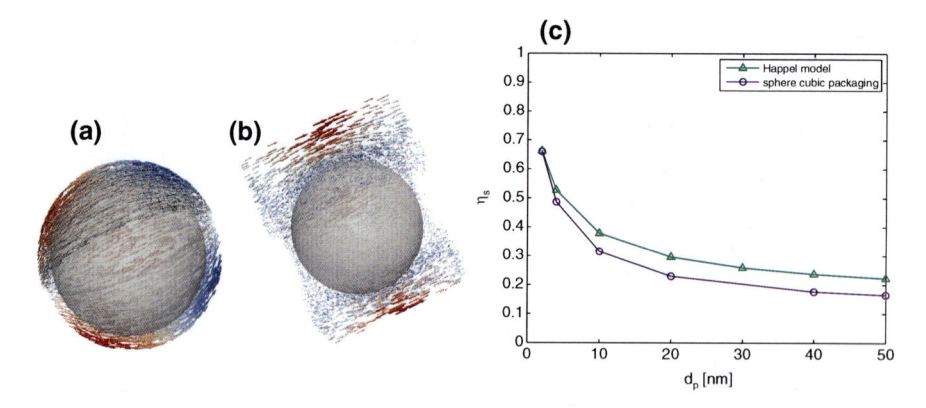

Fig. 6.8 Nanofluid motion around a spherical cell at fixed porosity $\epsilon = 0.6649$ in the case of Happel's model and our cubic packing. (**a**) Streamlines of fluid velocity in Happel's sphere-in-cell model; (**b**) streamlines for a spherical cell with cubic packing; (**c**) comparison of the trend of single cell absorption efficiency versus particle size in the two cases

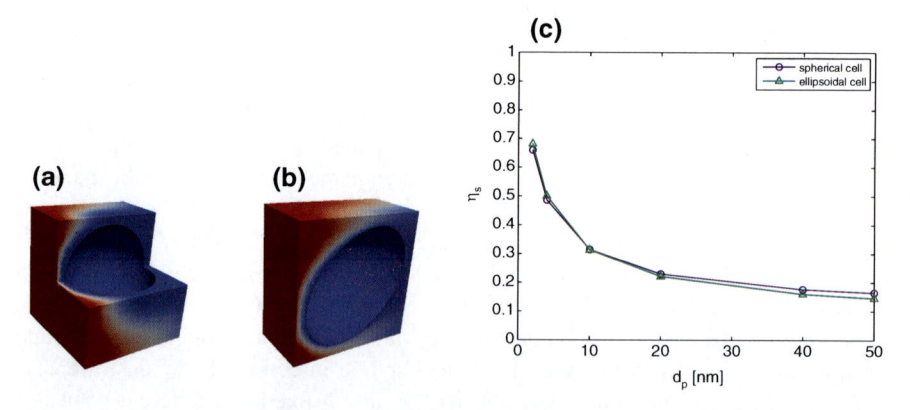

Fig. 6.9 Nanofluid motion around a single cell of different geometry at fixed porosity $\epsilon = 0.6649$. (**a**) Nanoparticles concentration around a sphere; (**b**) nanoparticles concentration around an ellipsoid; (**c**) comparison of the trend of single cell absorption efficiency versus particle size in the case of different cell geometries with fixed porosity $\epsilon = 0.6649$ and cubic packing; blue line represents spherical cell, green line ellipsoidal cell

6.2.4.3 Nanofluid Motion into a Pore/Layer

Let us now consider the nanofluid motion into either a sinusoidal pore or a sinusoidal layer. As in the previous analysis, we have first numerically computed the fluid velocity, i.e. by solving Eq. (6.7), and then we calculated the nanoparticle concentration through the solution of equation (6.9). In Fig. 6.10 we depict the single cell efficiency in the case of a sinusoidal pore and a sinusoidal layer at fixed porosity and fixed tortuosity τ, defined as the ratio between the amplitude a of the sinusoidal and the corresponding wavelength L. We find that the sinusoidal pore has

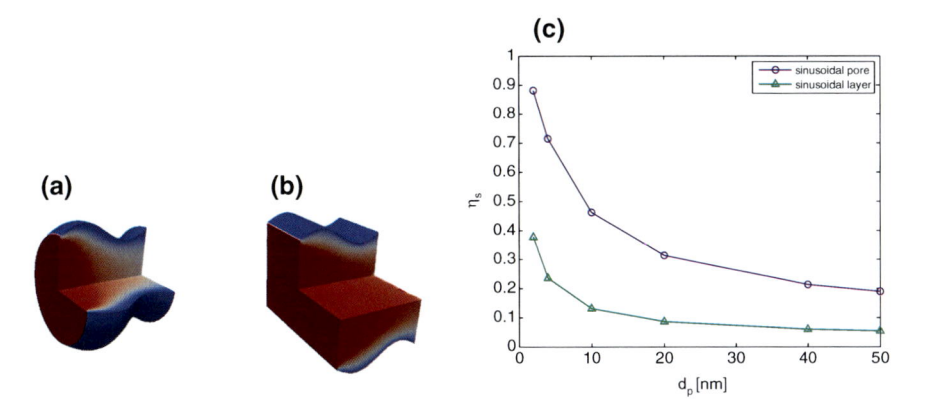

Fig. 6.10 Nanofluid motion into a pore of different geometry at fixed porosity $\epsilon = 0.6649$. (**a**) Sinusoidal pore; (**b**) sinusoidal layer; (**c**) trend of single cell absorption efficiency versus particle size in the two cases; blue line represents sinusoidal pore, green line sinusoidal layer

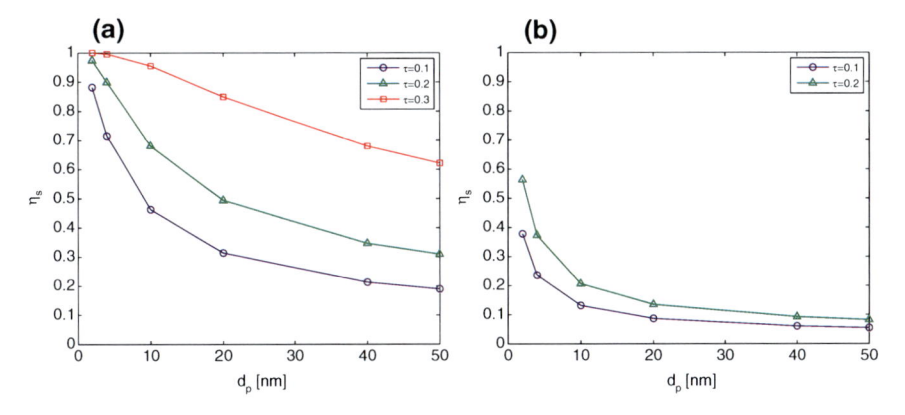

Fig. 6.11 Trend of single cell absorption efficiency versus particle size for the motion of the nanofluid into a pore with different tortuosity, in the case of (**a**) a sinusoidal pore, (**b**) a sinusoidal layer. $L = \overline{w} = 10$ μm and $\epsilon = 0.6649$

a bigger cell efficiency for all the particles sizes under consideration. The effect of tortuosity on the cell efficiency is shown in Fig. 6.11 for both the sinusoidal pore and the sinusoidal layer. We set $\tau = 0.1, 0.2$ for the sinusoidal layer and $\tau = 0.1, 0.2, 0.3$ for the sinusoidal pore: in both cases, the efficiency decreases for decreasing tortuosity.

6.3 A Macroscale Approach to Transport in Vascularized Tissues

Microvascular fluid dynamics plays a fundamental role in determining the efficiency of the drug delivery of nanoparticles. In this section we study the flow in the capillary network coupled with interstitial filtration, see Eq. (6.19), and transport of particles described by Eq. (6.20). As sketched in Fig. 6.12, flow in microvessels, interstitial flow and transport are coupled phenomena to be modeled by space-time dependent partial differential equations, which can be efficiently solved using advanced numerical technique, like the *embedded multiscale method*, developed in [15, 16] and later applied in [8, 9] for studying perfusion and drug delivery.

The geometrical configuration of the macroscale model is shown in Fig. 6.12: a tumor slab of R3230AC mammary carcinoma on a rat model, whose data are reported in [48] and made available as a result of the Microcirculation Physiome Project [43]. The tumor slab size is $220 \times 208 \times 92 \; 10^{-5}$ m and it contains 105 different segments with radii spanning from 5.5×10^{-6} m to 33.2×10^{-6} m. The variability in the radius introduces several technical difficulties in the mathematical model; for example the enforcement of mass conservation at each bifurcation of the network has not yet been addressed in the available computational solver. For this reason, we adopt here a constant radius $R = 7.64 \times 10^{-6}$ m that is the arithmetic average of the individual radii of each segment. A generalization of the flow model to varying capillary radius is presented in [41] and it will be used in forthcoming studies.

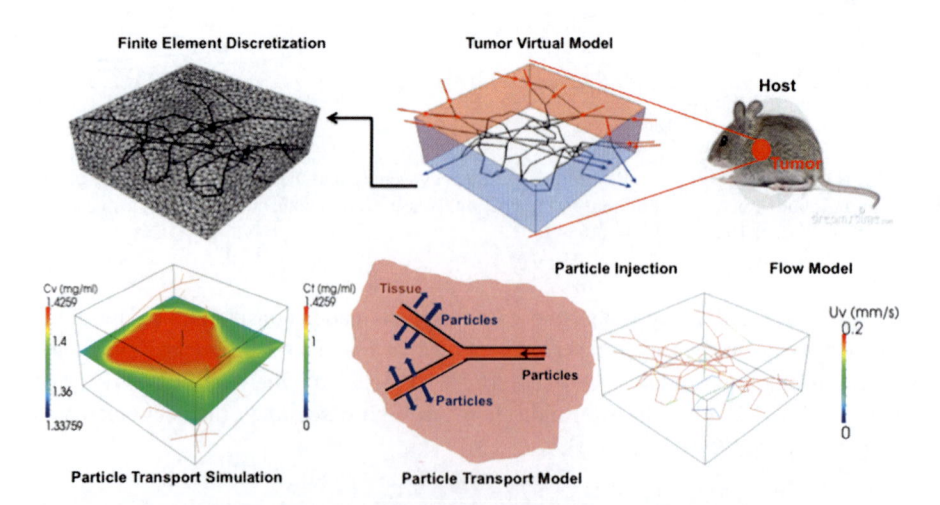

Fig. 6.12 Sketch of the VMN-based hyperthermia process, split into different phases: (1) manufacturing of the particles; (2) definition of the delivery protocol; (3) set up of the FEM computational model; (4) simulation of the VMN spatio-temporal distribution; (5) simulation of hyperthermia and study of the temperature maps

The computational model is decomposed into the microvascular network Λ and the surrounding malignant tissue Ω. The subscript v (vascular) will denote all the variables defined in the capillary domain, while the subscript t labels the tissue. Our computations simulate a protocol for HCT where the tumor slab is infused with a solution of particles. The injected particles enter the (virtual) tumor model through the inlets of the capillary network. These inlets are represented by the specific points of the capillaries intersecting with the chosen faces of the tumor slab.

The physical quantities of interest are the flow pressure p, the velocity $\mathbf{u}$ and the concentration of transported solutes c. All fields depend on time t and space, $\mathbf{x} \in \Omega$ being the spatial coordinates. The mathematical model stems from fundamental balance laws regulating the flow in the capillary bed, the extravasation of plasma and solutes and their transport in the interstitial tissue.

6.3.1 Governing Equations of Flow at the Macroscale

The flow model includes two components, i.e. the microcirculation and the flow in the interstitial tissue, coupled by interface conditions at the microvascular wall, behaving as a semipermeable membrane. The malignant interstitial volume is assumed to be an isotropic porous material, where the flow obey the Darcy's law.

Microcirculation is characterized by low Womersley and Reynolds numbers [4, 6, 50]. In these conditions, the Navier-Stokes equations reduce to steady Stokes flow. Under additional assumptions of (i) straight channels, (ii) rigid walls (iii) and constant radius, (iv) no-slip boundary conditions for the velocity, (v) absence of body forces such as gravity and inertial forces, Poiseuille flow can be exactly integrated, namely (6.19e). As discussed in [6, 21, 22, 49], although assuming Poiseuille flow considerably reduces the computational cost, some assumptions do not fully apply in the present context. Indeed, deviations must be small: if transmural flow occurs, it has to be small with respect to the axial component of the velocity; each vessel branch does not have to be straight, but only a small curvature is allowed. Improvements of the current flow model are in order: (a) for a better approximation of the extravasation effects; (b) to relieve constraints on small curvature and consequently allow to analyze configurations characterized by high tortuosity, which plays an important role on the average hydraulic conductivity of tumors, as discussed in [42]; (c) to improve the characterization of stresses on the wall, which in turn will affect particle adhesion.

The arc length coordinate along each capillary segment is represented by the symbol s, while $\boldsymbol{\lambda}$ denotes the reference vector that characterizes the segment orientation. Furthermore, we follow [5, 51] to model the lymphatic drainage.

The coupled mathematical model for microcirculation and flow in interstitial volume allows to find the pressure p_t, p_v and the velocity fields $\mathbf{u}_t$, $\mathbf{u}_v$ such that the flow problem is:

$$-\nabla \cdot \left(\frac{\kappa}{\mu}\nabla p_t\right) + L_p^{LF}\frac{S}{v}(p_t - p_L) - f_b(\bar{p}_t, p_v)\delta_\Lambda = 0, \quad \text{in } \Omega, \quad (6.19\text{a})$$

$$\mathbf{u}_t = -\frac{\kappa}{\mu}\nabla p_t, \quad \text{in } \Omega, \quad (6.19\text{b})$$

$$-\frac{\pi R^4}{8\mu}\frac{\partial^2 p_v}{\partial s^2} + f_b(\bar{p}_t, p_v) = 0, \quad s \in \Lambda, \quad (6.19\text{c})$$

$$f_b(\bar{p}_t, p_v) = 2\pi R L_p((p_v - \bar{p}_t) - \sigma^p(\pi_v^p - \pi_t^p)), \quad \text{in } \Lambda, \quad (6.19\text{d})$$

$$\mathbf{u}_v = -\frac{R^2}{8\mu}\frac{\partial p_v}{\partial s}\lambda, \quad s \in \Lambda, \quad (6.19\text{e})$$

$$-\frac{\kappa}{\mu}\nabla p_t \cdot \mathbf{n} = \beta_b(p_t - p_0), \quad \text{on } \partial\Omega. \quad (6.19\text{f})$$

where $\bar{p}_t$ represents an average of the interstitial pressure acting on the capillary surface, namely

$$\bar{p}_t(s) = \frac{1}{2\pi R}\int_0^{2\pi} p_t(s, \theta)R d\theta,$$

being θ the angular coordinate on the cylindrical surface representing the capillary wall.

We have imposed a pressure gradient along the capillary vessels. Since both the inflow and outflow of the capillary vessels are located on the outer edges of the slab, we enforced a known pressure p_{in} on two neighbor inlet, whereas outlet pressure p_{out} is assigned on the opposite faces. The pressure drop between inlets and outlets is calibrated according to the following argument. Given an estimate of the average blood velocity in the capillary circulation, equal to 0.1 mm/s according to the data provided by Intaglietta et al. [32], we use Poiseuille's law to calculate the corresponding pressure drop. More precisely, we have applied a fictitious model consisting of a straight rigid pipe of length $|\Lambda|$ and radius R, to calculate the pressure drop $p_{in} - p_{out}$ that corresponds to a velocity of 0.1 mm/s in the pipe. We have imposed the Robin-type boundary conditions (6.19f) for the blood flow in the interstitial volume. In this equation, p_0 stands for the far field pressure value, while β_b denotes an effective flow conductivity accounting for the tissue layers surrounding the tumor slab.

Let L_p be the hydraulic permeability of the vessel wall (see Table 6.2 for units and physiological values) and let $p_v - \bar{p}_t$ be the pressure difference between

Table 6.2 Parameters and data for mass and thermal transport

Symbol	Parameter	Units	Value	Source	Equation
L_p	Hydraulic permeability, capillary wall	(m^2 s)/kg	10^{-10}	[8]	(6.20b)
$L_p^{LF}\frac{S}{V}$	Effective permeability, Lymphatic Vessels	(mmHg hour)$^{-1}$	0.5	[8]	(6.20c)
c_{inj}	Inflow particle concentration	gr/m^3	1425.9	[8]	–
d	Edge length of particles	m	1×10^{-8}	[10]	–
m	Mass of particles	gr	8×10^{-18}	[10]	–
D_v	Vascular diffusivity of particles	m^2/s	9.0687×10^{-11}	[8]	(6.20a)
D_t	Interstitial diffusivity of particles	m^2/s	1.2955×10^{-11}	[8]	(6.20c)
P	Vascular permeability of particles	m/s	2×10^{-6}	[8]	(6.20b)

the vessels and the interstitial volume. Because of osmosis, the pressure drop across the capillary wall is a function of difference in chemical potential of the chemicals soluted in blood, [13, 24]: this determines the oncotic pressure jump $(\pi_v^p - \pi_t^p)$ modulated by the sieving coefficient σ^p. The oncotic pressure is defined as $\pi = R_g T c$, where c is the concentration of a given osmotic agent, R_g is the universal gas constant and T stands for the absolute temperature. The coefficient σ^p accounts for the difference of a semipermeable membrane compared to the case of ideal permeability (i.e., no resistance force on the molecules passing through the membrane). It spans from 0 to 1, where small values characterize ideal membranes, while larger values are typical of selective filters.

Proteins dissolved into the blood serum, and albumin in particular, are responsible for most of the oncotic pressure generated in the capillaries, which in physiological conditions is about 25 mmHg. To model this effect we set $\pi_v^p = 25$, $\pi_t^p = 0$ mmHg and $\sigma^p \simeq 1$, since proteins hardly leak through the capillary walls. As a result we obtain $\sigma^p (\pi_v^p - \pi_t^p) = 25$ mmHg (1 mmHg = 133.322 Pa). The oncotic effect generated by the injected VMNs is neglected in this work. Nevertheless, this is an open question to be explored in future studies, since albumin serum concentration is only 5–10 times bigger than the VMN systemic concentrations reached after injection.

In order to model the capillary phenotype typically observed in tumors, we increase the magnitude of the hydraulic permeability κ as in [5], such that the model will account of the well known enhanced permeability and retention effect (EPR). To balance leakage of arterial capillaries, venous and lymphatic systems absorb the fluid in excess. For the sake of generality, we include lymphatic drainage in the model, although the lymphatic system may be disfunctional in tumors. Following [5, 51], we model them as a distributed sink term in the interstitial volume. It is assumed that the volumetric flow rate due to lymphatic vessels, Φ^{LF}, is proportional to the pressure difference between the interstitium and the lymphatics, namely $\Phi^{LF}(p_t) = L_p^{LF} \frac{s}{v}(p_t - p_L)$, where L_p^{LF} is the hydraulic permeability of the lymphatic wall, s/v is the surface area of lymphatic vessels per unit volume of tissue and p_L is the hydrostatic pressure within the lymphatic channels.

6.3.2 Governing Equations of Mass Transport at the Macroscale

Mass transport in the capillary bed is modelled by means of advection-diffusion equations. As shown in [14], a one dimensional model for mass transport in the capillaries network can be derived starting from the 3D advection-diffusion problem. The coupled problem, accounting for transport of chemicals from the microvasculature to the interstitium, dictates the evolution of the concentrations c_v and c_t respectively.

In the interstitial tissue the particles or molecules are advected by the fluid and diffuse in all Ω. In addition they may be metabolised by the cells in the interstitial tissue. The distribution of solutes in the interstitial tissue is also affected by the lymphatic drainage. According to the assumptions at the basis of the flow model, the effect of lymphatic drainage on mass transport is represented as a distributed sink proportional to $L_p^{LF} \frac{S}{v}(p_t - p_L)c_t$.

Given this notation and assumptions, the mass transport model at the macroscale reads as follows:

$$\frac{\partial c_v}{\partial t} + \frac{\partial}{\partial s}\left((\mathbf{u}_v \cdot \boldsymbol{\lambda})c_v - D_v \frac{\partial c_v}{\partial s}\right) = -\frac{1}{\pi R^2} f_c(\overline{p}_t, p_v, \overline{c}_t, c_v), \qquad \text{in } \Lambda \times (0,t), \tag{6.20a}$$

$$f_c(\overline{p}_t, p_v, \overline{c}_t, c_v) = 2\pi R\big[L_p((p_v - p_t) - \sigma^p(\pi_v^p - \pi_t^p))c_v + P(c_v - c_t)\big], \qquad \text{in } \Lambda, \tag{6.20b}$$

$$\frac{\partial c_t}{\partial t} + \nabla \cdot (c_t \mathbf{u}_t - D_t \nabla c_t) + L_p^{LF} \frac{S}{v}(p_t - p_L)c_t + k_f c_t = f_c(\overline{p}_t, p_v, \overline{c}_t, c_v)\delta_\Lambda, \qquad \text{in } \Omega \times (0,t), \tag{6.20c}$$

$$(c_t \mathbf{u}_t - D_t \nabla c_t) \cdot \mathbf{n} = \beta_c c_t, \qquad \text{on } \partial\Omega \times (0,t) \tag{6.20d}$$

where D_v and D_t are the particle diffusivities, in the capillaries and the interstitium, respectively, assumed to be constant in each region. The rate of metabolization in the interstitium is denoted by k_f and, as an instance for many other parameters, can be calculated on the basis of the microscale model illustrated above. Unfortunately, simulations for a specific class of nanoparticles where the macroscopic model is informed by the microscale are not yet available but will be the result of work in progress based on the general framework proposed here.

We describe the capillary walls as *semipermeable membranes* allowing leakage of fluid and selective filtration of molecules. Again, the Kedem-Katchalsky equations represent a good model for these phenomena [24]. Then, under the assumption that capillaries can be modeled as cylindrical channels, the magnitude of the mass flux exchanged per unit length between the network of capillaries and the interstitial volume at each point of the capillary vessels is

$$f_c(\overline{p}_t, p_v, \overline{c}_t, c_v) = 2\pi R\big[(1 - \sigma)J_b(\overline{p}_t, p_v)c_v + P(c_v - \overline{c}_t)\big].$$

We posit that a constant concentration of particles, denoted by c_{inj}, is available in the blood flowing into the slab through the inflow sections of the vasculature. The particles are free to leave the system through the complementary outflow boundaries. At the initial time the vascular network and the tumor slab do not contain particles. For closing the transport problem, we model the layers of tissue surrounding the tumor sample by means of a condition that prescribes the flow resistance due to the outer layers of tissue, namely Eq. (6.20d).

6.3.3 Computational Solver

The discretization of problems (6.19) and (6.20) is performed by using the Finite Element Method. After partitioning the tumor and vasculature domains, Ω and Λ respectively, into elements, (see Fig. 6.12 top right panel showing a representative computational domain of only 32,624 tetrahedral elements) the solution of the governing equations are approximated with piecewise polynomial functions in the framework of the variational formulation. In particular, piecewise linear finite elements are used for all the unknowns, namely p_t, c_t and p_v, c_v, on a computational grid consisting of 49,655 grid points and 272,872 tetrahedral elements. Velocities $\mathbf{u}_t$, $\mathbf{u}_v$ are reconstructed in the post-processing phase using the pressure fields. We have adopted the GMRES method with incomplete-LU preconditioning to solve the algebraic systems following from the finite element discretization. The sensitivity of the results with respect to the mesh size has been tested and mesh independence is shown for grids finer than 257,109 elements.

The domains Ω and Λ feature heterogeneous dimensionality. The former is 3D, the latter is 1D. In order to model the natural leakage of capillaries, we apply the *embedded multiscale method* [8, 9, 15, 16], where the capillary bed is represented as a network of one-dimensional channels acting as concentrated sources of mass for the interstitial volume. The main advantage of the proposed scheme is that the computational grids required to approximate the equations on the capillary network and on the interstitial volume are completely independent. As a result, arbitrarily complex microvascular geometries can be studied with modest computational effort. From the standpoint of numerical approximation, the theoretical aspects of the method have been addressed in the works by D'Angelo [15, 16]. These algorithms have been implemented using GetFem++, a general C++ finite element library [47].

6.3.4 Numerical Simulations

A standard protocol for nano-based cancer treatment is not yet available. We have performed virtual experiments using data from previous studies that target nano-based tumor hyperthermia [34, 38, 54]. On the basis of these examples, we analyze a time interval of 60 min, where for the initial 40 min the tumor is supplied with a solution of particles, ideally provided by an intravascular injection of particles. The underlying assumption is that, for a small animal, the intravenous infusion of a nano-fluid generates an initial homogeneous concentration in the entire systemic circulation, which we denote as c_{inj}. In particular, we have chosen to run experiments targeting the reference value $c_{ref} = 1$ mg/ml because it matches the injected concentrations used in the experiments of [10]. We use the computational model to perform the following studies: (1) an analysis of average particle concentration time-course during injection whose results are reported in Fig. 6.13; (2) combined spatial maps of interstitial fluid pressure, concentration and temperature. In particular, Fig. 6.14 shows particle concentration field at 40 min., the time when particle injection is switched off.

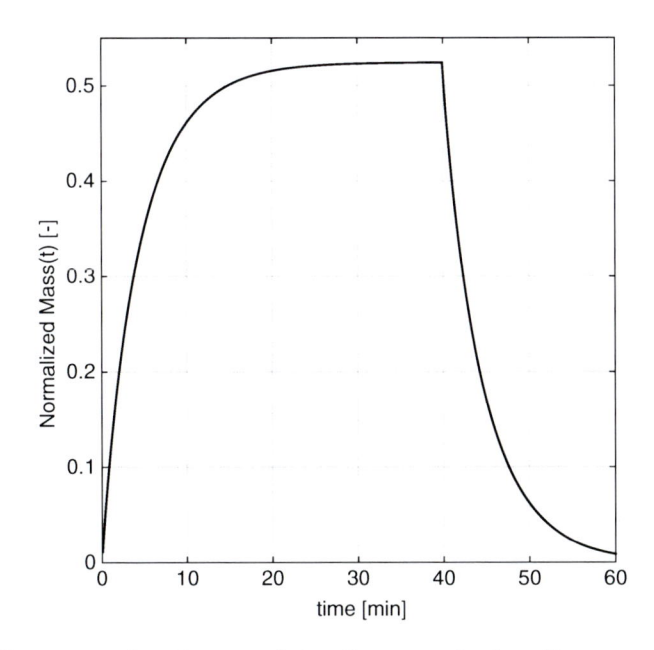

Fig. 6.13 Time-course of particle accumulation after renormalization with respect to the amount of particles injected for 40 min

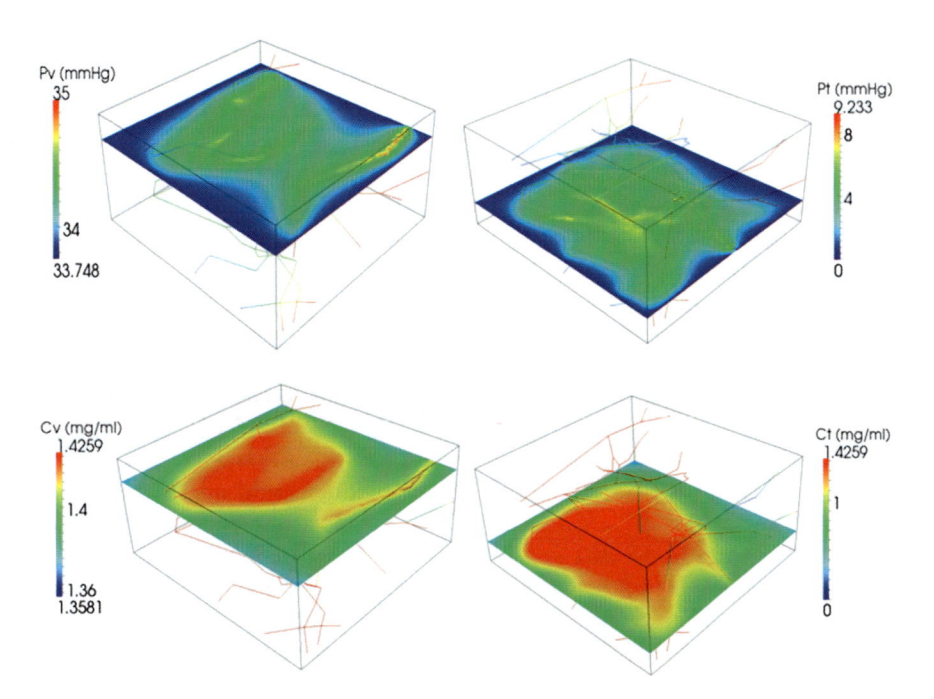

Fig. 6.14 Interstitial fluid pressure, p_t for high hydraulic permeability and absent lymphatic drainage $L_p^{LF} \frac{s}{v} = 0$ (top panel). Spatial distribution of c_t, c_v for particle delivery with constant target $c_{ref} = 1$ mg/ml (bottom panel)

In Fig. 6.13, the mass density of particles delivered to the tumor slab is shown. In these simulations the injected particle density has been set to match reference slab concentration of 1 mg/ml. Two characteristic traits of small-particle delivery emerge. Figure 6.13 shows the delivery efficiency, namely the mass of delivered particles per unit mass of delivered material. More precisely we plot the normalized mass as

$$\frac{\int_\Omega c_t d\mathbf{x} + \int_\Lambda \pi R^2 c_v ds}{\int_0^T \left(\pi R^2 c_{inj} \mathbf{u}_{in} \cdot \boldsymbol{\lambda}\right) dt}. \qquad (6.21)$$

We observe that more than 50% of the injected particles are absorbed by the tumor slab, because of their ability to extravasate and diffuse within the interstitial volume. However, for similar reasons, small particles suffer from small residence times, as demonstrated by the quick drop of particle concentration and temperature after 40 min. In other words, the particle concentration significantly drops as soon as particle injection is switched off.

For the same simulations, particle concentration and temperature fields are shown in Fig. 6.14. For a highly vascularized small tumor (about 0.5 mm width) the particle distribution among the tissue is rather uniform.

6.4 Conclusions and Future Perspectives

In this work we have proposed a multiscale mathematical model for the description of the diffusion, transport and absorption of nanoparticles in living tissues. In particular, using a bottom-up approach, the upscaling technique introduced allows to determine the absorption rate of nanoparticles in a continuous macroscopic model, resulting in a one-way coupling between micro- and macro-scales.

We have addressed microscale simulations first. We have implemented a discrete and a continuous microscopic models in two representative geometries, mimicking both the absorption rate in a regular packing of cellular aggregates and the nanoparticle extravasation through the sinusoidal surface of a pore/layer. The corresponding microscopic models account for the presence of hydrodynamic retardations and driving forces, namely the lift and the Wan der Waals interaction forces, and they have been solved using Kinetic Monte Carlo and the Finite Volume methods, respectively. The results of the microscopic numerical simulations allowed to analyze the effects of the different physical factors acting on a nanoparticle on the cell efficiency. In the case of a cellular absorption, we have proposed a more realistic cubic packing for both a spherical and a ellipsoidal unit cell, showing that the absorption efficiency deviates significantly from the previous results obtained using idealized Happel's model: therefore, we remark that a numerical approach is required to properly identify the local absorption parameter since the analytical simplification can drive to results that are off by as much as 20%.

We have also found that both the hydrodynamic retardations and the van der Waals interaction forces influence the single cell efficiency, although with opposite effects. Similar results are found also for the pore/layer geometry mimicking the extravasation process, where the cell efficiency is strongly dependent of the surface tortuosity.

In conclusion, we have demonstrated that the transport of nanoparticles in living tissues is strongly affected by the geometrical and the multi-physical factors at the microscale. Therefore, the results of this work push towards the development of a more accurate microscopic models in order to improve the level of approximation at which the transport properties across the scales must be investigated. Although we have validated the upscaling technique of the proposed multiscale approach, future refinements would take into account other relevant physical phenomena. For example, other driving forces at the microscale should be considered, such as the electrostatic potential deriving from surface functionalization [1] or from the protein corona which inevitably forms on a nanoparticle's surface in a biological medium [37].

At the macroscale level, we have outlined new mathematical model designed to investigate the coupled flow and mass transfer at the level of the whole tumor. Basic balance and constitutive laws have been adopted to simulate the interactions among the main compartments of the tumor tissue. The main contribution of the computational framework is that flow and mass transfer in the capillary and interstitial medium are coupled for simulations performed using a microvasculature configuration based on physiological data. The simulations suggest that network topology and particle distribution along the microvasculature are the key factors in particle delivery approach, especially in the cases characterized by hindered particle extravasation.

One of the main limitations of the macroscopic approach is the difficulty of determining the model parameters. Table 6.2 lists the most significant ones, such as the hydraulic permeability of the capillary walls, the particle diffusivity tensor in the blood stream and in the tissue matrix, the vascular permeability of particles and the deposition rate coefficient. All these coefficients are usually estimated from available literature sources based on experimental data or phenomenological models. The corresponding values are affected by large uncertainties. For any specific family of nanoparticles, a mechanistic approach, as the one developed here at the microscale, would bring huge benefits in reducing the margins of error with respect of the parameter values. Here, we have outlined the application of this approach to the estimate of the deposition rate coefficient, but the general methodology find ubiquitous applications to improve the reliability of macroscale simulations, which root in this preliminary investigation and will be surely at the core of future research efforts.

Acknowledgements We would thank Stefania Lunardi for the help in collecting the information presented in the introductory part and her contribution in the numerical simulations with the LKMC technique for the microscopic model and Michele Pollini for the help in conducting the FE continuum simulations for the microscopic model. Funding by the AIRC grant MFAG

17412 is gratefully acknowledged. DA, PC and MT are members of Gruppo Nazionale di Fisica Matematica (GNFM) of the Istituto Nazionale di Alta Matematica (INdAM). CdF and PZ gratefully acknowledge support by the Gruppo Nazionale di Calcolo Scientifico (GNCS) of INdAM.

References

1. Albanese A, Tang P, Chan W (2012) The effect of nanoparticle size, shape, and surface chemistry on biological systems. Ann Rev Biomed Eng 14(1):1–16
2. Arnold D, Brezzi F, Fortin M (1984) A stable finite element for the stokes equations. Calcolo 21(4):337–344
3. Banerjee RK, van Osdol WW, Bungay PM, Sung C, Dedrick RL (2001) Finite element model of antibody penetration in a prevascular tumor nodule embedded in normal tissue. J Control Release 74(1):193–202
4. Baxter L, Jain R (1989) Transport of fluid and macromolecules in tumors. I. role of interstitial pressure and convection. Microvasc Res 37(1):77–104
5. Baxter L, Jain R (1990) Transport of fluid and macromolecules in tumors II. role of heterogeneous perfusion and lymphatics. Microvasc Res 40(2):246–263
6. Blake T, Gross J (1982) Analysis of coupled intra- and extraluminal flows for single and multiple capillaries. Math Biosci 59(2):173–206
7. Carmeliet P, Jain R (2000) Angiogenesis in cancer and other diseases. Nature 407(6801): 249–257
8. Cattaneo L, Zunino P (2014) A computational model of drug delivery through microcirculation to compare different tumor treatments. Int J Numer Methods Biomed Eng 30(11):1347–1371
9. Cattaneo L, Zunino P (2014) Computational models for fluid exchange between microcirculation and tissue interstitium. Netw Heterog Media 9(1):135–159
10. Cervadoro A, Giverso C, Pande R, Sarangi S, Preziosi L, Wosik J, Brazdeikis A, Decuzzi P (2013) Design maps for the hyperthermic treatment of tumors with superparamagnetic nanoparticles. PLoS One 8(2):e57332
11. Cherukat P, McLaughlin JB (1994) The inertial lift on a rigid sphere in a linear shear flow field near a flat wall. J Fluid Mech 263:1–18
12. Chou CY, Huang CK, Lu KW, Horng TL, Lin WL (2013) Investigation of the spatiotemporal responses of nanoparticles in tumor tissues with a small-scale mathematical model. PloS One 8(4):e59135
13. Curry F (1984) Mechanics and thermodynamics of transcapillary exchange. In: Handbook of physiology, chap 8, pp 309–374. American Physiological Society, Bethesda
14. D'Angelo C (2007) Multiscale modeling of metabolism and transport phenomena in living tissues. PhD thesis, Politecnico di Milano
15. D'Angelo C (2012) Finite element approximation of elliptic problems with dirac measure terms in weighted spaces: applications to one- and three-dimensional coupled problems. SIAM J Numer Anal 50(1):194–215
16. D'Angelo C, Quarteroni A (2008) On the coupling of 1D and 3D diffusion-reaction equations. application to tissue perfusion problems. Math Models Methods Appl Sci 18(8):1481–1504
17. Decuzzi P, Ferrari M (2007) The role of specific and non-specific interactions in receptor-mediated endocytosis of nanoparticles. Biomaterials 28(18):2915–2922
18. Decuzzi P, Ferrari M (2008) The receptor-mediated endocytosis of nonspherical particles. Biophys J 94(10):3790–3797
19. Elimelech M (1994) Particle deposition on ideal collectors from dilute flowing suspensions: mathematical formulation, numerical solution, and simulations. Sep Technol 4(4):186–212
20. Flamm MH, Diamond SL, Sinno T (2009) Lattice kinetic monte carlo simulations of convective-diffusive systems. J Chem Phys (130):094904

21. Fleischman G, Secomb T, Gross J (1986) The interaction of extravascular pressure fields and fluid exchange in capillary networks. Math Biosci 82(2):141–151
22. Flieschman G, Secomb T, Gross J (1986) Effect of extravascular pressure gradients on capillary fluid exchange. Math Biosci 81(2):145–164
23. Florence AT (2012) "targeting" nanoparticles: the constraints of physical laws and physical barriers. J Control Release 164(2):115–124
24. Friedman M (2008) Principles and models of biological transport. Springer, New York
25. Gao Y, Li M, Chen B, Shen Z, Guo P, Wientjes MG, Au JLS (2013) Predictive models of diffusive nanoparticle transport in 3-dimensional tumor cell spheroids. AAPS J 15(3):816 831
26. Goldman A, Cox R, Brenner H (1967) Slow viscous motion of a sphere parallel to a plane wall-II Couette flow. Chem Eng Sci 22(4):653–660
27. Goldman A, Cox RG, Brenner H (1967) Slow viscous motion of a sphere parallel to a plane wall-I motion through a quiescent fluid. Chem Eng Sci 22(4):637–651
28. Goodman TT, Chen J, Matveev K, Pun SH (2008) Spatio-temporal modeling of nanoparticle delivery to multicellular tumor spheroids. Biotechnol Bioeng 101(2):388–399
29. Graff CP, Wittrup KD (2003) Theoretical analysis of antibody targeting of tumor spheroids importance of dosage for penetration, and affinity for retention. Cancer Res 63(6):1288–1296
30. Hanahan D, Weinberg R (2000) The hallmarks of cancer. Cell 100(1):57–70
31. Hicks K, Pruijn F, Secomb T, Hay M, Hsu R, Brown J, Denny W, Dewhirst M, Wilson W (2006) Use of three-dimensional tissue cultures to model extravascular transport and predict in vivo activity of hypoxia-targeted anticancer drugs. J Natl Cancer Inst 98(16):1118–1128
32. Intaglietta M, Silverman N, Tompkins W (1975) Capillary flow velocity measurements in vivo and in situ by television methods. Microvasc Res 10(2):165–179
33. Israelachvili JN (2011) Intermolecular and surface forces: revised third edition. Academic, New York
34. Johannsen M, Gneveckow U, Eckelt L, Feussner A, Waldöfnerr N, Scholz R, Deger S, Wust P, Loening SA, Jordan A (2005) Clinical hyperthermia of prostate cancer using magnetic nanoparticles: presentation of a new interstitial technique. Int J Hyperth 21(7):637–647
35. Logg A, Wells GN (2010) Dolfin: automated finite element computing. ACM Trans Math Softw 37(2):20:1–20:28
36. Lunardi S (2014) Simulazione microscala dell'efficienza di assorbimento di nanoparticelle per trasporto di farmaco. Master's thesis, Politecnico di Milano
37. Monopoli M, Bombelli FB, Dawson K (2011) Nanobiotechnology: nanoparticle coronas take shape. Nature Nanotech 6:11–12
38. Mornet S, Vasseur S, Grasset F, Duguet E (2004) Magnetic nanoparticle design for medical diagnosis and therapy. J Mater Chem 14(14):2161–2175
39. Nelson KE, Ginn TR (2005) Colloid filtration theory and the happel sphere-in-cell model revisited with direct numerical simulation of colloids. Langmuir 21(6):2173–2184
40. Norris E, King JR, Byrne HM (2006) Modelling the response of spatially structured tumours to chemotherapy: drug kinetics. Math Comput Model 43(7):820–837
41. Notaro D, Cattaneo L, Formaggia L, Scotti A, Zunino P (2016) A mixed finite element method for modeling the fluid exchange between microcirculation and tissue interstitium. In: Ventura G, Benvenuti E (eds) Advances in discretization methods: discontinuities, virtual elements, fictitious domain methods. Springer, Cham, pp 3–25. https://doi.org/10.1007/978-3-319-41246-7_1
42. Penta R, Ambrosi D (2015) The role of the microvascular tortuosity in tumor transport phenomena. J Theoret Biol 364:80–97
43. Popel A, Greene A, Ellis C, Ley K, Skalak T, Tonellato P (1998) The microcirculation physiome project. Ann Biomed Eng 26:911–913
44. Putti M, Cordes C (1998) Finite element approximation of the diffusion operator on tetrahedra. SIAM J Sci Comput 19(4):1154–1168
45. Quintard M, Whitaker S (1995) Aerosol filtration: an analysis using the method of volume averaging. J Aerosol Sci 26(8):1227–1255

46. Rajagopalan R, Tien C (1976) Trajectory analysis of deep-bed filtration with the sphere-in-cell porous media model. AIChE J 22(3):523–533
47. Renard Y, Pommier J (2012). http://download.gna.org/getfem/html/homepage/
48. Secomb T, Hsu R, Braun R, Ross J, Gross J, Dewhirst M (1998) Theoretical simulation of oxygen transport to tumors by three-dimensional networks of microvessels. Adv Exp Med Biol 454:629–634
49. Secomb T, Park RH EYH, Dewhirst M (2004) Green's function methods for analysis of oxygen delivery to tissue by microvascular networks. Ann Biomed Eng 32(11):1519–1529
50. Skalak R, Keller S, Secomb T (1981) Mechanics of blood flow. J Biomech Eng 103(2):102–115
51. Soltani M, Chen P (2013) Numerical modeling of interstitial fluid flow coupled with blood flow through a remodeled solid tumor microvascular network. PLoS One 8(6). https://doi.org/10.1371/journal.pone.0067023
52. Su D, Ma R, Salloum M, Zhu L (2010) Multi-scale study of nanoparticle transport and deposition in tissues during an injection process. Med Biol Eng Comput 48(9):853–863
53. Taffetani M, de Falco C, Penta R, Ambrosi D, Ciarletta P (2014) Biomechanical modelling in nanomedicine: multiscale approaches and future challenges. Arch Appl Mech 84(9–11):1627–1645
54. Terentyuk G, Maslyakova G, Suleymanova L, Khlebtsov N, Khlebtsov B, Akchurin G, Maksimova I, Tuchin V (2009) Laser-induced tissue hyperthermia mediated by gold nanoparticles: toward cancer phototherapy. J Biomed Opt 14(2):021016
55. Tufenkji N, Elimelech M (2004) Correlation equation for predicting single-collector efficiency in physicochemical filtration in saturated porous media. Environ Sci Technol 38(2):529–536
56. van Osdol W, Fujimori K, Weinstein JN (1991) An analysis of monoclonal antibody distribution in microscopic tumor nodules: consequences of a"binding site barrier". Cancer Res 51(18):4776–4784
57. Waite CL, Roth CM (2011) Binding and transport of PAMAM-RGD in a tumor spheroid model: the effect of RGD targeting ligand density. Biotechnol Bioeng 108(12):2999–3008
58. Ward JP, King JR (2003) Mathematical modelling of drug transport in tumour multicell spheroids and monolayer cultures. Math Biosci 181(2):177–207
59. Ying CT, Wang J, Lamm RJ, Kamei DT (2013) Mathematical modeling of vesicle drug delivery systems 2 targeted vesicle interactions with cells, tumors, and the body. J Lab Autom 18(1):46–62

Chapter 7
A Continuum Model of Skeletal Muscle Tissue with Loss of Activation

Giulia Giantesio and Alessandro Musesti

7.1 Introduction

Skeletal muscle tissue is one of the main components of the human body, being about 40% of its total mass. Its principal role is the production of *force*, which supports the body and becomes *movement* by acting on bones. The mechanism by which a muscle produces force is called *activation*.

Skeletal muscle tissue is a highly ordered hierarchical structure. The cells of the tissue are the muscular fibers, having a length up to several centimeters; they are organized in fascicles, where every fiber is multiply connected to nerve axons, which drive the activation of the tissue. Connective tissue, which is essentially isotropic, fills the spaces among the fibers. Every fiber contains a concatenation of millions of sarcomeres, which are the fundamental unit of the muscle. With a length of some micrometers, a sarcomere is composed by chains of proteins, mainly actin and myosin, which can slide on each other. This sliding movement produces the contraction of the sarcomere and, ultimately, the contraction of the whole muscle and the production of force and movement.

The aim of this chapter is to propose a mathematical model of skeletal muscle tissue with a reduced activation, which is typical of a geriatric syndrome named *sarcopenia* [16]. About 30 years ago, the term sarcopenia (from Greek *sarx* or flesh and *penia* or loss) has been introduced in order to describe the age-related decrease of muscle mass and performance. Sarcopenia has since then been defined as the loss of skeletal muscle mass and strength that occurs with advancing age, which in turn affects balance, gait and overall ability to perform even the simple

---

G. Giantesio · A. Musesti (✉)
Dipartimento di Matematica e Fisica, Università Cattolica del Sacro Cuore, via Musei 41,
25121 Brescia, Italy
e-mail: giulia.giantesio@unicatt.it; alessandro.musesti@unicatt.it

tasks of daily living such as rising from a chair or climbing steps. According to [6], sarcopenia affects more than 50 millions people today and it will affect more than 200 millions in the next 40 years. There is still no generally accepted test for its diagnosis and many efforts are made nowadays by the medical community to better understand this syndrome. Therefore it is desirable to build a mathematical model of muscle tissue affected by sarcopenia. However, to the best of our knowledge, in the biomathematical literature the topic of loss of activation has never been addressed.

In order to use the valuable tools of Continuum Mechanics, during the last decades the skeletal muscle tissue has been often modeled as a continuum material [4, 5, 9, 11], which is usually assumed to be transversely isotropic and incompressible. The former assumption is motivated by the alignment of the muscular fibers, while the latter is ensured by the high water content of the tissue (about 75% of the total volume). Moreover, in view of some experimental tests, the material is assumed to be nonlinear and viscoelastic. Focusing our attention only on the steady properties of the tissue, here we neglect the viscous effects and we set in the framework of hyperelasticity.

In the model that we propose, there are three constitutive prescriptions: one for the hyperelastic energy when the tissue is not active (*passive energy*), one for the activation and one for the loss of performance. As far as the passive part is concerned, we assume an exponential stress response of the material, which is customary in biological tissues. The particular form that we choose, being a slight simplification of the one proposed in [9], has the advantage of being polyconvex and coercive, giving mathematical soundness to the model and stability to the numerical simulations.

A recent and very promising way to describe the activation is the *active strain* approach, where the extra energy produced by the activation mechanism is encoded in a multiplicative decomposition of the deformation gradient in an elastic and an active part (see Sect. 7.2.2). Unlike the classical *active stress* approach, in which the active part of the stress is modeled in a pure phenomenological way and a new term has to be added to the passive energy, the active strain method does not change the form of the elastic energy, keeping in particular all its mathematical properties. Moreover, at least in the case of skeletal muscles, the active strain approach seems to be more adherent to the physiology of the tissue, in the sense that at the molecular level the production of force is actually given by a deformation of the material, thanks to the contraction of the sarcomeres. The active part of the deformation gradient is a mathematical representation of such a contraction. The multiplicative decomposition of the deformation gradient has been applied to an active striated muscle in [12, 18]. However, this decomposition involves only a part of the whole elastic energy, which is written as the sum of two terms for the case of a fiber-reinforced material. As far as we know, the active strain approach has never been previously applied to the whole elastic energy of a skeletal muscle tissue. As a drawback, the active strain approach can be a source of some technical difficulties; for instance, in our case fitting the model on the experimental data is not so simple, see Eq. (7.13).

Furthermore, we consider the loss of performance, which is one of the novelties of our model. Unfortunately, there are no experimental data on the elastic properties of a sarcopenic muscle tissue, at least to our knowledge; hence we adopt the naive strategy of reducing the active part of the stress (which is the difference between the stress of the material with and without activation) by a given percentage, represented by the damage parameter d (see Sect. 7.3.2). In this way, there is a single parameter in the model which concisely accounts for any effect of the disease.

The proposed model can be numerically implemented using finite element methods. In Sect. 7.4 we present some results obtained using FEniCS, an open source collection of Python libraries. Actually, we consider a cylindrical geometry with radial symmetry, so that the numerical domain is two-dimensional and the computational cost is reduced. As far as the boundary conditions are concerned, we prescribe the displacement on the bases of the cylinder and let the lateral surface traction-free. Such simulations show that the experimental results of [10] on the passive and active stress-strain healthy curves, obtained *in vivo* from a tetanized tibialis anterior of a rat, can be well reproduced by our model. Further, the behavior of the tissue when d increases is analyzed. An ongoing task is to perform a finite element implementation of the model when generic loads are applied, and to consider a realistic three-dimensional muscle mesh. We are now developing a truly hyperelastic model, where the expression of the stress takes into account also the dependence of the activation on the deformation gradient. Actually, in this chapter the stress is computed as the derivative of the hyperelastic energy keeping the active part of the deformation gradient fixed.

In the future, it will be very interesting to find some connections between the damage percentage (the parameter d) and other physiological quantities, such as the mass of the muscular tissue or the neuronal activity. Another important topic will be the application of some homogenization techniques in order to deduce an improved constitutive equation for the skeletal muscle starting from its microstructure.

7.2 Constitutive Model

Skeletal muscle tissue is characterized by densely packed muscle fibers, which are arranged in fascicles. Filling the spaces between the fibers and fascicles, connective tissue surrounds the muscle and it is responsible of the elastic recoil of the muscle to elongation. Besides a large amount of water, the fibers themselves contain titin, actin and myosin filaments. The latter two sliding elements form the actual contractile component of the muscle, which is called *sarcomere*. Since the fibers locally follow a predominant unidirectional alignment, transverse isotropy with respect to that main direction can be assumed. We hence begin by modelling the skeletal muscle tissue as a transversely isotropic nonlinear hyperelastic material with principal direction **m**, which follows the alignment of the muscle fibers.

7.2.1 Passive Model

Let $\mathbf{F}$ denote the deformation gradient tensor, $\mathbf{C} = \mathbf{F}^T\mathbf{F}$ the right Cauchy-Green tensor and $\mathbf{M} = \mathbf{m} \otimes \mathbf{m}$ the so called *structural tensor*. If Ω denotes the reference configuration occupied by the muscle, we describe its passive behavior by choosing a hyperelastic strain energy function

$$\int_\Omega W(\mathbf{C}) dV,$$

where the strain energy density is of the form

$$W(\mathbf{C}) = \frac{\mu}{4} \left\{ \frac{1}{\alpha} \left[e^{\alpha(I_p - 1)} - 1 \right] + K_p - 1 \right\}, \qquad (7.1)$$

with

$$I_p = \frac{w_0}{3} \operatorname{tr}(\mathbf{C}) + (1 - w_0) \operatorname{tr}(\mathbf{CM}), \quad K_p = \frac{w_0}{3} \operatorname{tr}(\mathbf{C}^{-1}) + (1 - w_0) \operatorname{tr}(\mathbf{C}^{-1}\mathbf{M}).$$

Here μ is an elastic parameter and α and w_0 are positive dimensionless material parameters. The generalized invariants I_p and K_p are given by a weighted combination of the isotropic and anisotropic components; in particular, w_0 measures the ratio of isotropic tissue constituents and $1 - w_0$ that of muscle fibers. Moreover, the term $\operatorname{tr}(\mathbf{CM})$ represents the squared stretch in the direction of the muscle fiber and is thus associated with longitudinal fiber properties, while the term $\operatorname{tr}(\mathbf{C}^{-1}\mathbf{M})$ describes the change of the squared cross-sectional area of a surface element which is normal to the direction $\mathbf{m}$ in the reference configuration and thus relates to the transverse behavior of the material [8, 20].

One of the mathematical features of the energy density (7.1) is that it is polyconvex and coercive [7, 20], hence the equilibrium problem with mixed boundary conditions is well posed.

We remark that $\mathbf{C}$ is the identity tensor $\mathbf{I}$ in the reference configuration, so that $I_p = K_p = 1$, i.e. we have the energy- and stress-free state of the passive muscle tissue (see [7]).

The high content of water is responsible of the nearly incompressible behavior which is experimentally reported for muscle fibers, so that we can assume

$$\det \mathbf{C} = 1. \qquad (7.2)$$

As is customary in hyperelasticity, the first Piola-Kirchhoff stress tensor, known as *nominal stress tensor*, can be directly computed by differentiating the strain energy function:

$$\mathbf{P} = \frac{\partial W}{\partial \mathbf{F}} - p\mathbf{F}^{-T} = 2\mathbf{F}\frac{\partial W}{\partial \mathbf{C}} - p\mathbf{F}^{-T} = \qquad (7.3)$$

$$= \frac{\mu}{2}\mathbf{F}\left\{e^{\alpha(I_p - 1)}\left[\frac{w_0}{3}\mathbf{I} + (1-w_0)\mathbf{M}\right] - \mathbf{C}^{-1}\left[\frac{w_0}{3}\mathbf{I} + (1-w_0)\mathbf{M}\right]\mathbf{C}^{-1}\right\} - p\mathbf{F}^{-T},$$

where p is a Lagrange multiplier associated with the hydrostatic pressure which results from the incompressibility constraint (7.2).

The material parameters of the model can be obtained from real data. More precisely, concerning the elastic parameter μ, we use the value given in [9], while the other two parameters have been obtained by least squares optimization using the experimental data by Hawkins and Bey [10] about the stretch response of a tetanized *tibialis anterior* of a rat (see Fig. 7.1). In Table 7.1 we furnish the values of the parameters. Figure 7.2 shows the anisotropic behavior of the model.

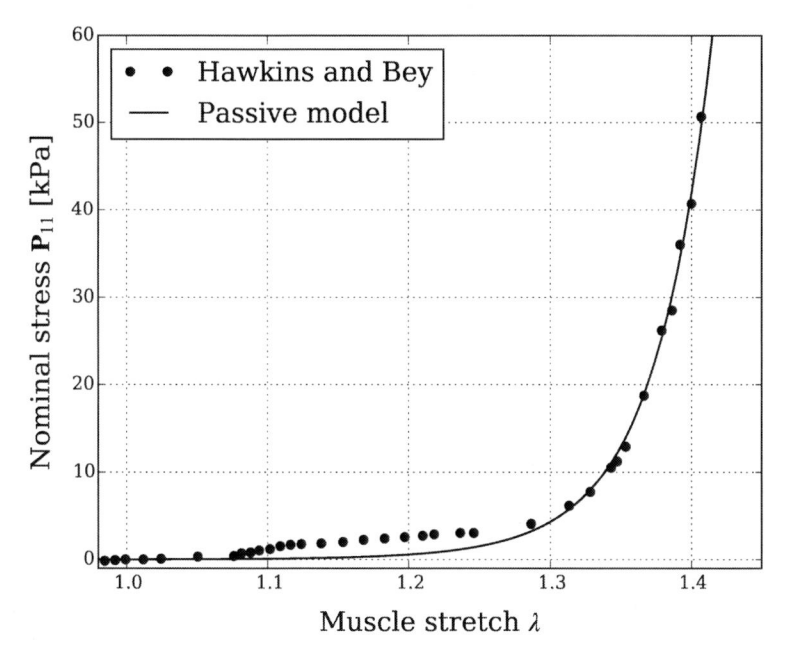

Fig. 7.1 Comparison of the passive model in uniaxial tension with the experimental data of a rat tibialis anterior muscle reported in [10]

Table 7.1 Material parameters of the passive model

μ [kPa]	α [–]	w_0 [–]
0.1599	19.35	0.7335

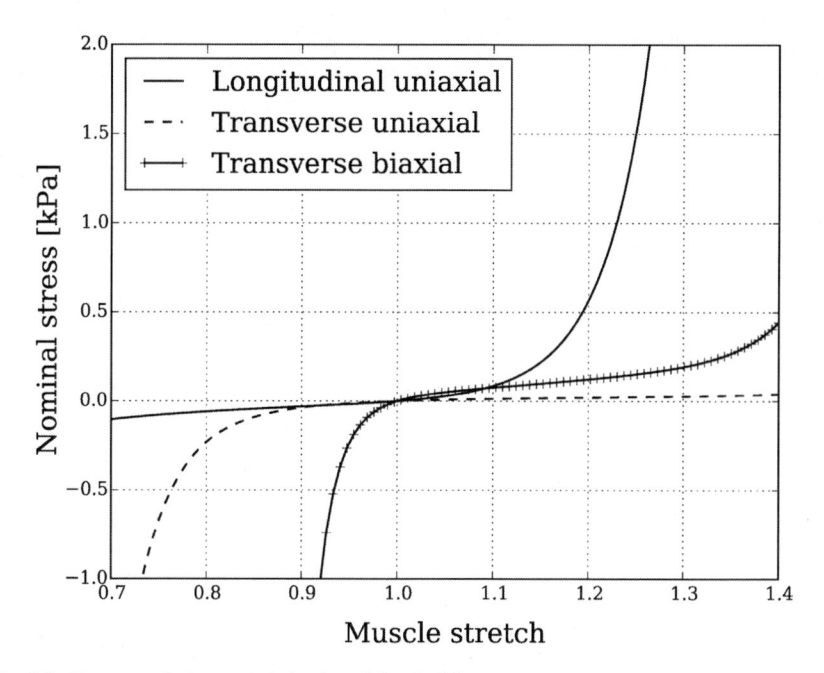

Fig. 7.2 Transversely isotropic behavior of the model

We remark that the strain energy function (7.1) is a slight simplification of the one proposed by Ehret, Böl and Itskov in [9]:

$$W_{\text{EBI}}(\mathbf{C}) = \frac{\mu}{4}\left\{\frac{1}{\alpha}\left[e^{\alpha(I_p-1)} - 1\right] + \frac{1}{\beta}\left[e^{\beta(K_p-1)} - 1\right]\right\}, \tag{7.4}$$

where $\alpha = 19.69, \beta = 1.190, w_0 = 0.7388$. Actually, our simplification consists in linearizing the term related to K_p, which describes the transverse behavior. This is motivated by the fact that the parameter β is much smaller than α. In Fig. 7.3 we can see the comparison between the nominal stress in the direction of the stretch of the two models when the muscle fibers are elongated in their direction.

7.2.2 Active Model

One of the main features of the skeletal muscle tissue is its ability of being voluntarily activated. Skeletal muscles are activated through electrical impulses from motor nerves; the activation triggers a chemical reaction between the actin and myosin filaments which produces a sliding of the molecular chains, causing a contraction of the muscle fibers.

7 A Continuum Model of Skeletal Muscle Tissue with Loss of Activation

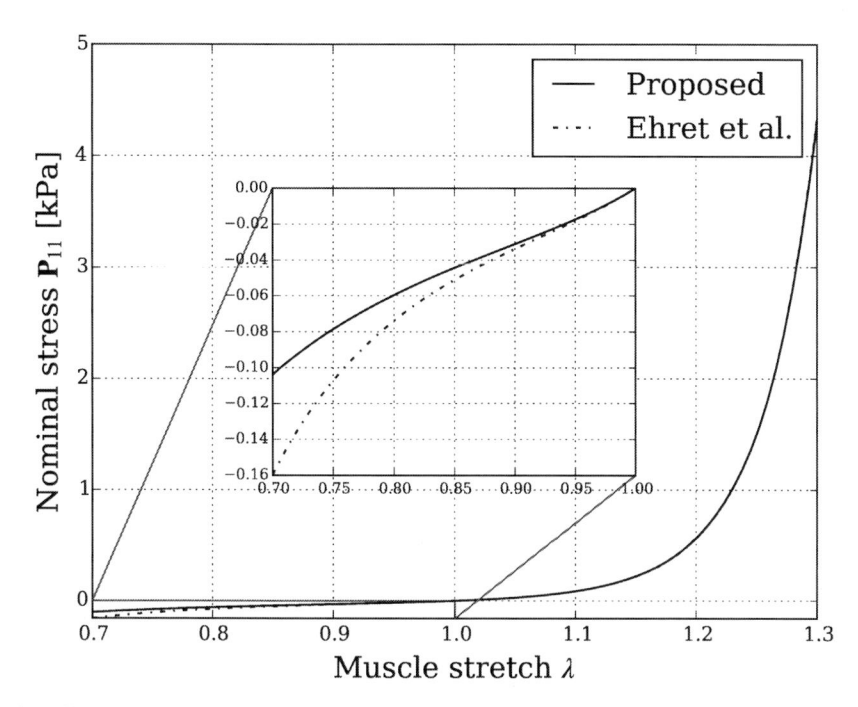

Fig. 7.3 Comparison between the passive stress here proposed and the one studied in [9] during uniaxial tension along the fibers

During the last decades, many authors tried to mathematically model the process of activation, mainly with two different approaches (for a review see [2]). The most famous approach followed in the literature is called *active stress* and it consists in adding an extra term to the stress, which accounts for the contribution given by the activation (see for example [3, 11, 15]). However, this is an *ad hoc* method, usually not related to the sliding movement of the filaments in the sarcomeres, which is the main mechanism of contraction at the mesoscale.

More recently, the *active strain* approach was proposed by Taber and Perucchio [21] in order to describe the activation of the cardiac tissue, following previous theories of growth and morphogenesis, as well as several models of plasticity. The method for soft living tissues is explained in [17]. Differently from the active stress approach, this method does not change the form of the strain energy function; rather, it assumes that only a part of the deformation gradient, obtained by a multiplicative decomposition, is responsible for the store of elastic energy. This method is related to the biological meaning of activation and can be reasonably adopted also in our case. To the best of our knowledge, the active strain approach has never been followed for the skeletal muscle tissue in literature.

We begin by rewriting the deformation gradient as $\mathbf{F} = \mathbf{F}_e \mathbf{F}_a$, where $\mathbf{F}_e$ is the elastic part and $\mathbf{F}_a$ describes the active contribution (see Fig. 7.4). The active strain $\mathbf{F}_a$ represents a change of the reference volume elements due to the contraction

Fig. 7.4 Pictorial view of the active strain approach

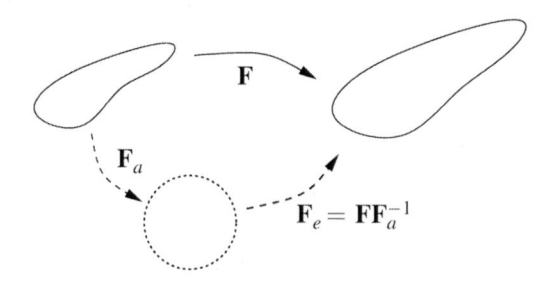

of the sarcomeres, so that it does not contribute to the elastic energy. A reference volume element, distorted by $\mathbf{F}_a$, needs a further deformation $\mathbf{F}_e$ to match the actual volume element, which accommodates both the external forces and the active contraction. Notice that neither $\mathbf{F}_a$ nor $\mathbf{F}_e$ need to be the gradients of some displacement, that is, it is not necessary that they fulfill the compatibility condition curl $\mathbf{F}_a = 0$ or curl $\mathbf{F}_e = 0$.

The volume elements are modified by the internal active forces without changing the elastic energy, hence the strain energy function of the activated material has to be computed using $\mathbf{C}_e = \mathbf{F}_e^T \mathbf{F}_e$ and taking into account $\mathbf{F}_e = \mathbf{F}\mathbf{F}_a^{-1}$. If $\mathbf{F}_a = \mathrm{grad}\,\chi_a$ for some displacement χ_a, then from Fig. 7.4 by a change of variables it is easy to see that

$$\int_{\chi_a(\Omega)} W(\mathbf{C}_e)d\widehat{V} = \int_{\Omega} W(\mathbf{F}_a^{-T}\mathbf{C}\mathbf{F}_a^{-1})(\det \mathbf{F}_a)dV.$$

The right-hand side of the previous equation is well defined also when $\mathbf{F}_a$ does not come from a global displacement, and it describes the strain energy of the active body. We then obtain the modified hyperelastic energy density

$$\widehat{W}(\mathbf{C}) = (\det \mathbf{F}_a)W(\mathbf{C}_e) = (\det \mathbf{F}_a)W(\mathbf{F}_a^{-T}\mathbf{C}\mathbf{F}_a^{-1}).$$

We now have to model the active part $\mathbf{F}_a$. Since the activation of the muscle consists in a contraction along the fibers, we choose

$$\mathbf{F}_a = \mathbf{I} - \gamma \mathbf{m} \otimes \mathbf{m}, \tag{7.5}$$

where $0 \leq \gamma < 1$ is a dimensionless parameter representing the relative contraction of activated fibers ($\gamma = 0$ meaning no activation). Then the modified strain energy density becomes

$$\widehat{W}(\mathbf{C}) = (1-\gamma)W(\mathbf{C}_e) = (1-\gamma)\frac{\mu}{4}\left\{\frac{1}{\alpha}\left[e^{\alpha(I_e-1)} - 1\right] + K_e - 1\right\}, \tag{7.6}$$

$$I_e = \frac{w_0}{3}\mathrm{tr}(\mathbf{C}_e) + (1-w_0)\,\mathrm{tr}(\mathbf{C}_e\mathbf{M}), \quad K_e = \frac{w_0}{3}\mathrm{tr}(\mathbf{C}_e^{-1}) + (1-w_0)\,\mathrm{tr}(\mathbf{C}_e^{-1}\mathbf{M}).$$

7 A Continuum Model of Skeletal Muscle Tissue with Loss of Activation

The corresponding first Piola-Kirchhoff stress tensor is given by

$$\widehat{\mathbf{P}} = \det \mathbf{F}_a \frac{\partial W}{\partial \mathbf{F}_e} \mathbf{F}_a^{-1} - \widehat{p} \mathbf{F}^{-T} = \tag{7.7}$$

$$= \frac{\mu}{2}(1-\gamma)\mathbf{F}_e \left\{ e^{\alpha(I_e-1)} \left[\frac{w_0}{3}\mathbf{I} + (1-w_0)\mathbf{M} \right] - \mathbf{C}_e^{-1}\left[\frac{w_0}{3}\mathbf{I} + (1-w_0)\mathbf{M}\right]\mathbf{C}_e^{-1}\right\} \mathbf{F}_a^{-1}$$

$$-\widehat{p}\mathbf{F}^{-T},$$

where $\widehat{p}$ accounts for the incompressibility constraint $\det \mathbf{C} = 1$. Notice that, since the activation (7.5) does not preserve volume and the material has to be globally incompressible, one has that $\det \mathbf{C}_e \neq 1$, so that the material is elastically compressible. As far as the strain energy density is concerned, a factor $(1-\gamma)$ appears in (7.6) which keeps into account the compressibility of $\mathbf{F}_a$. It would be interesting to study also other kinds of passive energies, involving the quantity $\det \mathbf{C}$, in order to better describe the elastic compressibility of the material.

In Fig. 7.5 we represent, for several values of the parameter γ, the stress-strain curve for a uniaxial tension along the fibers. If the muscle is activated ($\gamma > 0$), then (the absolute value of) the stress increases with γ and the value of the stretch such that the stress is zero becomes less than one.

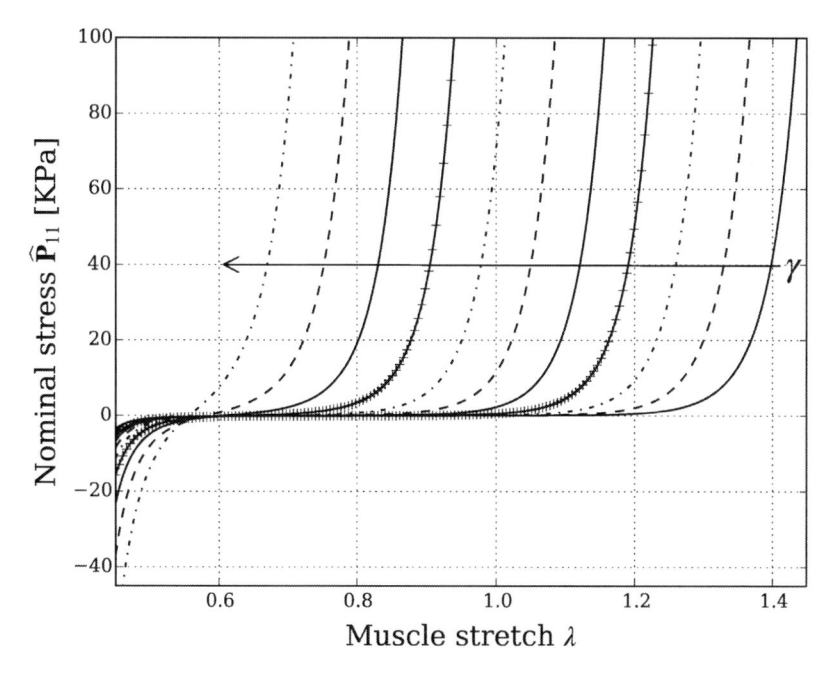

Fig. 7.5 Stress-stretch curves in uniaxial tension for several values of γ from 0 to 0.4

7.3 Modelling the Activation on Experimental Data

The activation parameter γ, which was assumed constant in the previous section, in fact usually depends on the deformation gradient. In typical experiments on a tetanized skeletal muscle it is apparent that the contraction of the fibers due to activation varies with their stretch, reaching a maximum value and then decreasing. Figure 7.6 shows the qualitative relation between the elongation and the developed stress. This section will be devoted to taking into account this phenomenon. Specifically, the expression of γ will be determined matching an experiment-based relation between stress and strain with our model (7.7).

In order to find the relation between stress and strain, the experiments *in vivo* are usually performed in two steps. First, the stress-strain curve is obtained without any activation (*passive curve*). Second, by an electrical stimulus the muscle is isometrically kept in a tetanized state and the *total stress-strain curve* is plotted. The last curve, which is qualitatively represented in Fig. 7.6, depends on the reciprocal position of actin and myosin chains. By taking the difference of the two curves one can obtain the *active curve*, describing the amount of stress due to activation. It is useful to find a mathematical expression of such a curve, in order to take into account the experimental behavior of the active contraction. This issue has already been addressed in several papers, see e.g. [3, 4, 9, 11, 13, 22, 23].

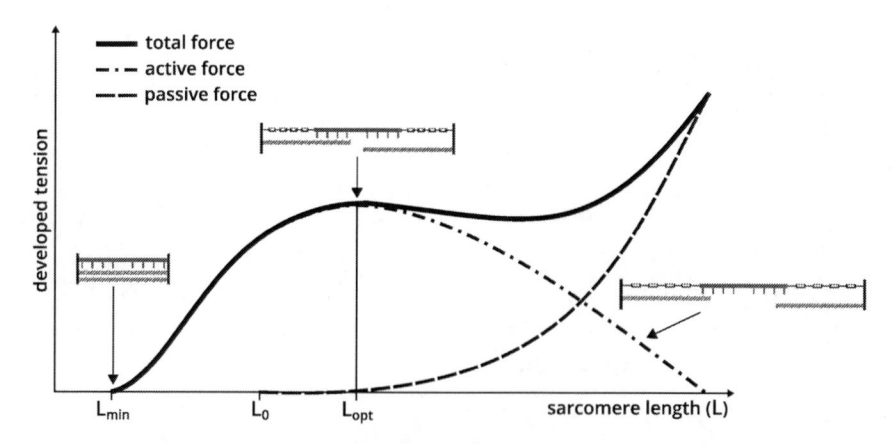

Fig. 7.6 Length-tension relationship of a sarcomere. Here we denote by L_{opt} the length at which the sarcomere produces the maximum force in isometric experiments, by L_0 the rest length and by L_{min} the minimal length of the sarcomere (fully activated)

7 A Continuum Model of Skeletal Muscle Tissue with Loss of Activation

Denoting with λ the ratio between the current length of the muscle and its original length, we assume the active curve to be of the form

$$P_{act}(\lambda) = \begin{cases} P_{opt} \exp\left[-k\dfrac{(\lambda^2 - \lambda_{opt}^2)^2}{\lambda - \lambda_{min}}\right] & \text{if } \lambda > \lambda_{min}, \\ 0 & \text{otherwise,} \end{cases} \qquad (7.8)$$

where λ_{min} is the minimum stretch value after which the activation starts (i.e. the lower bound for the stretch at which the myofilaments begin to overlap) and k is merely a fitting parameter. The coordinates (λ_{opt}, P_{opt}) identify the position of the maximum of the curve. As it is explained in [9], the value of P_{opt} takes into account some information at the mesoscale level, such as the number of activated motor units and the interstimulus interval; according to the literature [9, 11], it is set at $P_{opt} = 73.52$ kPa. The numerical values of the other three parameters, deduced through least squares optimization on the data reported in [10], are given in Table 7.2. The expression (7.8) has the advantage of describing the asymmetry between the ascending and descending branches of the active curve obtained in [10]. Indeed, even if the asymmetry is not so evident in their curve, due to the fact that there are only few data on the descending branch, it is a typical feature of several experimentally measured sarcomere length-force relation. Moreover, as one can easily see in Fig. 7.7, the convex behavior of the data nearby λ_{min} is well fitted.

7.3.1 The Activation Parameter γ as a Function of the Elongation

Now our aim is to obtain $P_{act}(\lambda)$ given in (7.8) from the model described in Sect. 7.2.2. In order to reach our purpose, we have to model the activation parameter γ as a function of the stretch.

As in the experiments of Hawkins and Bey [10], let us consider a uniaxial simple tension along the fibers. For simplicity, we assume that the fibers follow the direction $\mathbf{m} = \mathbf{e}_1$. Since the skeletal muscle tissue is modeled as an incompressible

Table 7.2 Material parameters of the active model

λ_{min} [–]	λ_{opt} [–]	k [–]	P_{opt} [kPa]
0.6243	1.1704	0.4342	73.52

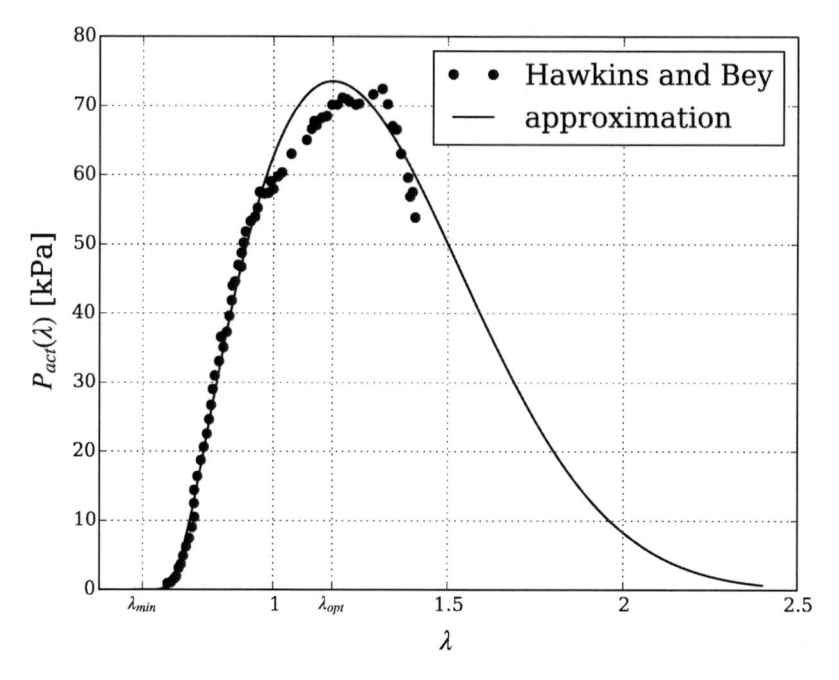

Fig. 7.7 Plot of the active curve (7.8) with the parameters reported in Table 7.2 together with the representation of the experimental data given in [10]

transversely isotropic material, the general form of the deformation gradient **F** is given by

$$\mathbf{F} = \begin{pmatrix} \lambda & 0 & 0 \\ 0 & \frac{1}{\sqrt{\lambda}} & 0 \\ 0 & 0 & \frac{1}{\sqrt{\lambda}} \end{pmatrix}.$$

Then using the notation introduced in Sect. 7.2.2, one has

$$\mathbf{C}_e = \begin{pmatrix} \frac{\lambda^2}{(1-\gamma)^2} & 0 & 0 \\ 0 & \frac{1}{\lambda} & 0 \\ 0 & 0 & \frac{1}{\lambda} \end{pmatrix},$$

$$I_e = \frac{w_0}{3}\left[\frac{\lambda^2}{(1-\gamma)^2} + \frac{2}{\lambda}\right] + (1-w_0)\frac{\lambda^2}{(1-\gamma)^2},$$

$$K_e = \frac{w_0}{3}\left[\frac{(1-\gamma)^2}{\lambda^2} + 2\lambda\right] + (1-w_0)\frac{(1-\gamma)^2}{\lambda^2}.$$

7 A Continuum Model of Skeletal Muscle Tissue with Loss of Activation

In this case, it is convenient to look at the strain energy as a function of the stretch λ and the activation parameter γ:

$$\widehat{W}(\lambda, \gamma) = (1-\gamma)W(\lambda, \gamma) = (1-\gamma)\frac{\mu}{4}\left\{\frac{1}{\alpha}\left[e^{\alpha(I_e-1)} - 1\right] + K_e - 1\right\}. \quad (7.9)$$

Then the nominal stress along the fiber direction is given by

$$P_{tot}(\lambda, \gamma) := \frac{\partial \widehat{W}}{\partial \lambda} = (1-\gamma)\frac{\mu}{4}\left[I'_e e^{\alpha(I_e-1)} + K'_e\right], \quad (7.10)$$

where

$$I'_e = \frac{\partial I_e}{\partial \lambda} = 2\frac{w_0}{3}\left[\frac{\lambda}{(1-\gamma)^2} - \frac{1}{\lambda^2}\right] + 2(1-w_0)\frac{\lambda}{(1-\gamma)^2},$$

$$K'_e = \frac{\partial K_e}{\partial \lambda} = 2\frac{w_0}{3}\left[-\frac{(1-\gamma)^2}{\lambda^3} + 1\right] - 2(1-w_0)\frac{(1-\gamma)^2}{\lambda^3}.$$

We can get the passive stress by setting $\gamma = 0$:

$$P_{pas}(\lambda) := P_{tot}(\lambda, 0) = \frac{\mu}{2}\left\{\left[\left(1 - \frac{2}{3}w_0\right)\lambda - \frac{w_0}{3}\frac{1}{\lambda^2}\right]e^{\alpha\left[\left(1-\frac{2}{3}w_0\right)\lambda^2 + \frac{w_0}{3}\frac{2}{\lambda} - 1\right]}\right.$$

$$\left. - \left(1 - \frac{2}{3}w_0\right)\frac{1}{\lambda^3} + \frac{w_0}{3}\right\}. \quad (7.11)$$

We remark that the values of P_{tot} and P_{pas} can also be obtained by computing the first component of the stress given by (7.7) and (7.3) after finding the hydrostatic pressure from the conditions $\widehat{P}_{22} = \widehat{P}_{33} = P_{22} = P_{33} = 0$ (traction-free lateral surface).

Our aim is to find the value of γ such that

$$P_{tot}(\lambda, \gamma) = P_{act}(\lambda) + P_{pas}(\lambda), \quad (7.12)$$

where $P_{act}(\lambda)$ is given by (7.8). Unfortunately, this leads to an equation for γ which cannot be explicitly solved:

$$(1-\gamma)\left\{\left[\left(1-\frac{2}{3}w_0\right)\frac{\lambda}{(1-\gamma)^2} - \frac{w_0}{3}\frac{1}{\lambda^2}\right]e^{\alpha\left[\left(1-\frac{2}{3}w_0\right)\frac{\lambda^2}{(1-\gamma)^2} + \frac{w_0}{3}\frac{2}{\lambda} - 1\right]}\right.$$

$$\left. + \frac{w_0}{3} - \left(1 - \frac{2}{3}w_0\right)\frac{(1-\gamma)^2}{\lambda^3}\right\} = \frac{2}{\mu}\left[P_{act}(\lambda) + P_{pas}(\lambda)\right]. \quad (7.13)$$

However one can employ standard numerical methods and plot the solution. Figure 7.8a, which is obtain by a bisection method, shows γ as a function of λ. We remark that γ vanishes before λ_{min}, indeed in this region there is no difference between total and passive stress. The corresponding behavior of the stresses is plotted in Fig. 7.8b, which is very similar to the representative plot of Fig. 7.6.

We emphasize that the previous model is not strictly hyperelastic, since in the expression of the stress (7.7) the derivative of γ with respect to $\mathbf{F}$ has been neglected. We are now working on a truly hyperelastic model, which can be useful for some numerical implementations.

7.3.2 Loss of Activation

We now want to describe from a mathematical point of view the loss of performance of a skeletal muscle tissue. As we have already explained in the Introduction, this is one of the main effects of *sarcopenia*, which is a typical syndrome of advanced age.

In [14, 24] it is remarked that aging is associated with changes in muscle mass, composition, activation and material properties. In sarcopenic muscle, there is a loss of motor units *via* denervation and a net conversion in slow fibers, with a resulting loss in muscle power. Hence, the loss of performance of a sarcopenic muscle can be described as a weakening of the activation of the fibers.

Unfortunately, as far as we know, there are no experimental data describing a uniaxial simple tension along the fibers of a sarcopenic muscle. For this reason, we try to describe the loss of activation by a parameter d which lowers the curve $P_{act}(\lambda)$ given by (7.8). The parameter d describes the percentage of *disease* or *damage*: if $d = 0$, then the muscle is healthy. In order to get our aim, we multiply the function $P_{act}(\lambda)$ by the factor $1 - d$, as one can see in Fig. 7.9. Notice that such a choice can be overly simple: for instance, it implies that the maximum is always attained at λ_{opt}, even if there is no experimental evidence of that. However, the presence of d allows to describe, at least qualitatively, the loss of performance of a muscle, which is one of the goals of our model.

7.4 Numerical Validation

Finally, we simulate numerically the contraction and the elongation of a slab of skeletal muscle tissue represented by a cylinder. We assume radial symmetry, so that the mesh is a rectangle. The ends of the cylinder are assumed to remain perpendicular to the axial direction. The rectangle is modeled by the hyperelastic model presented in the previous sections. The active contractile fibers are aligned along the length of the rectangle, which coincides with $\mathbf{e}_1$. The passive and active material parameters are given in Tables 7.1 and 7.2, respectively. Concerning the boundary conditions, the cylinder is fixed at one end and elongated to a given length,

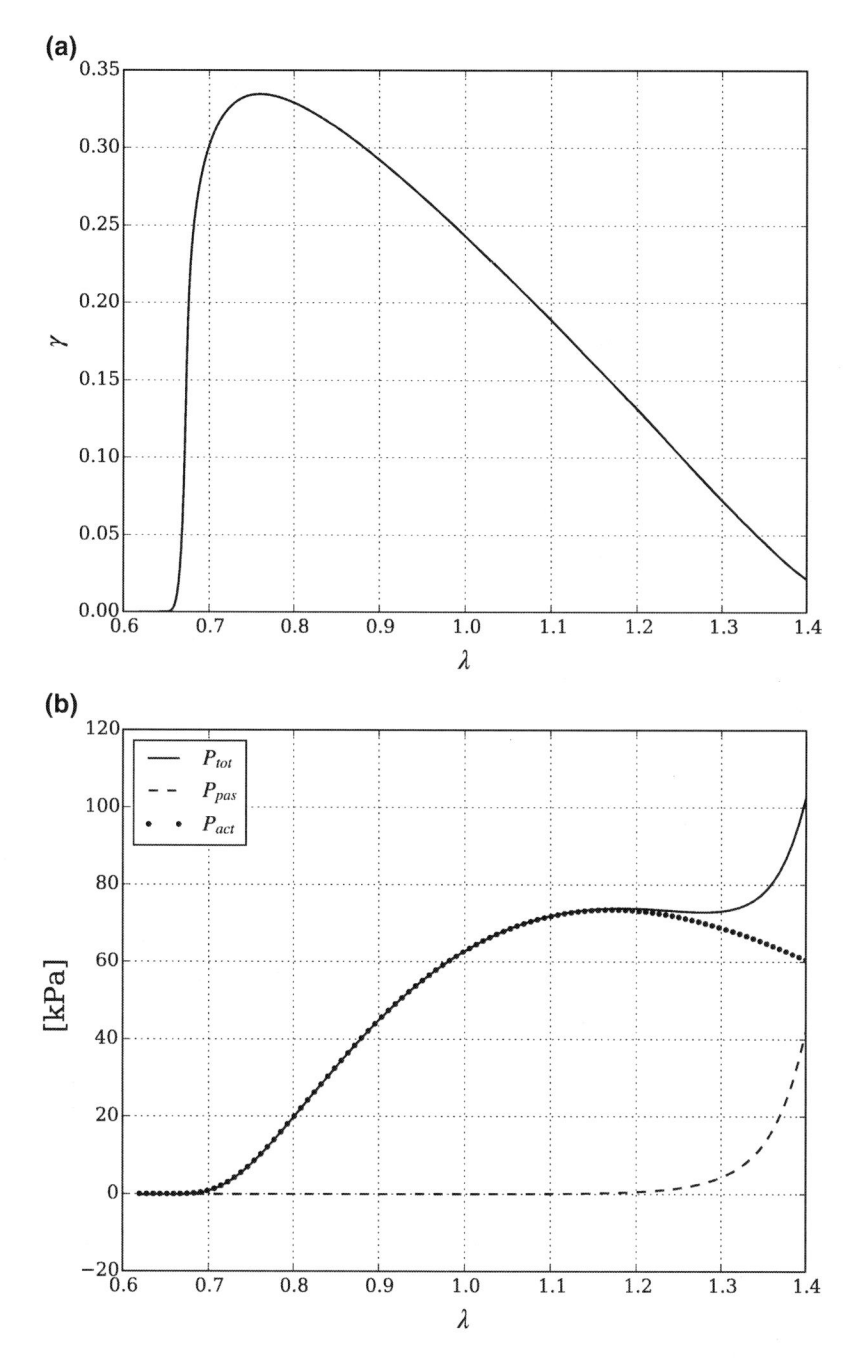

Fig. 7.8 (a) the first figure shows the behavior of γ when λ varies: the corresponding plots of P_{tot} and P_{act} are given in the second figure (**b**), together with P_{pas}

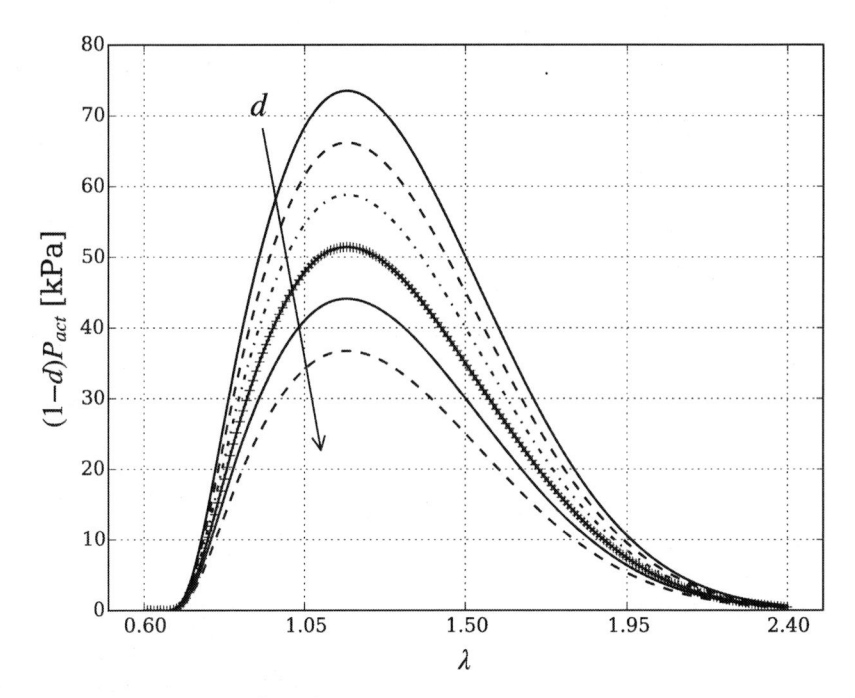

Fig. 7.9 Plot of $(1-d)P_{act}(\lambda)$ vs λ when d varies from 0 to 0.5

in order to recreate the situation of the experiments reported in [10]. The lateral surface is assumed to be tension-free.

The analysis is performed by using the computing environment FEniCS. The FEniCS Project [1] is a collection of numerical software, supported by a set of novel algorithms and techniques, aimed at the automated solution of differential equations using finite element methods.

As it is explained in Sect. 7.3.1, one of the main features of our model is the dependence of the activation parameter γ on the stretch λ. The function $\gamma(\lambda)$ solves the implicit equation (7.13), which ensures that the corresponding stress curves fit the experimental data. However, even if this equation can be solved using numerical methods, it is interesting to find an explicit function in order to analyze qualitatively the active model and to run the simulations in FEniCS. Moreover, the explicit function $\gamma(\lambda)$ has to be very precise, since a slight error on γ deeply affects the behavior of the total stress. Hence, it is reasonable to relate the expression of γ to the material parameters and the quantities involved in (7.13). An idea is to isolate the exponential in (7.13) and to express its exponent by a first step approximation

7 A Continuum Model of Skeletal Muscle Tissue with Loss of Activation

of a fixed-point method. We then obtain the following expression of γ:

$$\gamma(\lambda) = \begin{cases} a\left[\sqrt{\dfrac{1-\frac{2}{3}w_0}{\frac{g(\lambda_{min})}{\alpha}+\frac{w_0}{3\lambda_{min}}}}\lambda_{min} - \sqrt{\dfrac{1-\frac{2}{3}w_0}{\frac{g(\lambda)}{\alpha}+\frac{w_0}{3\lambda}}}\lambda\right] & \text{if } \lambda > \lambda_{min}, \\ 0 & \text{otherwise}, \end{cases} \quad (7.14)$$

$$g(\lambda) = \ln\alpha + \alpha\left(1 - \frac{w_0}{\lambda}\right) - \frac{1}{2}\ln\left(\frac{1-\frac{2}{3}w_0}{\frac{1}{\alpha}+\frac{w_0}{3\lambda}}\right)$$

$$+ \ln\left\{b\frac{2}{\mu}\left[P_{act}(\lambda) + P_{pas}(\lambda)\right] + \sqrt{\frac{1-\frac{2}{3}w_0}{\frac{1}{\alpha}+\frac{w_0}{3\lambda}}}\left[\frac{\left(1-\frac{2}{3}w_0\right)^2}{\frac{1}{\alpha}+\frac{w_0}{3\lambda}} - \lambda\frac{w_0}{3}\right]\right\},$$

where a and b are dimensionless fitting parameters: a is related to the magnitude of γ, while b acts on the curves (7.8) and (7.11), which are the terms of the equation not depending on γ. Performing a least square optimization on the resulting P_{act}, one gets $a = 1.0133$ and $b = 0.2050$.

Figure 7.10 shows the plot of the function $\gamma(\lambda)$ given in (7.14) in comparison to the numerical solution of equation (7.13) obtained by a bisection method. Notice

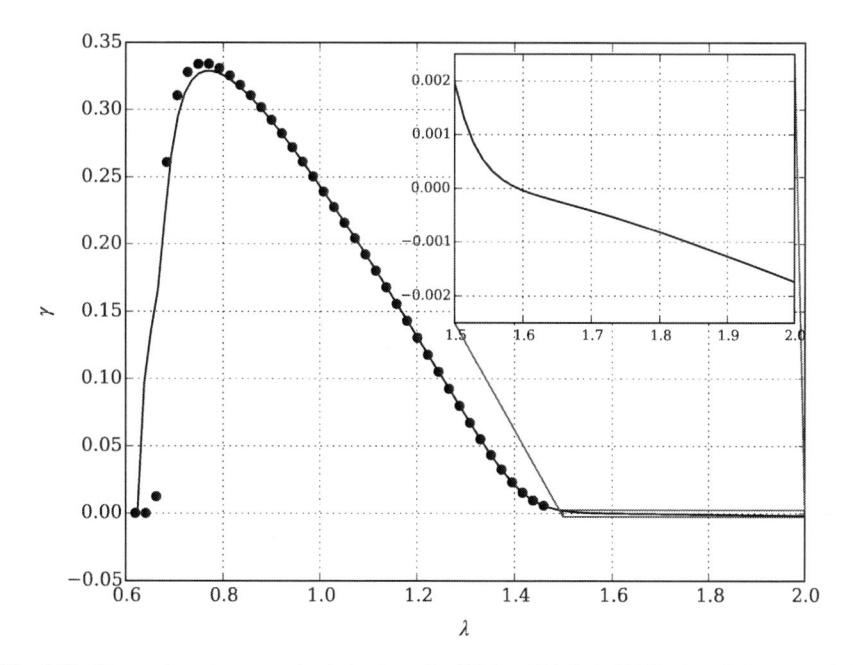

Fig. 7.10 Comparison between the behavior of $\gamma(\lambda)$ in (7.14) (solid line) and the numerical solution of equation (7.13) (dotted line)

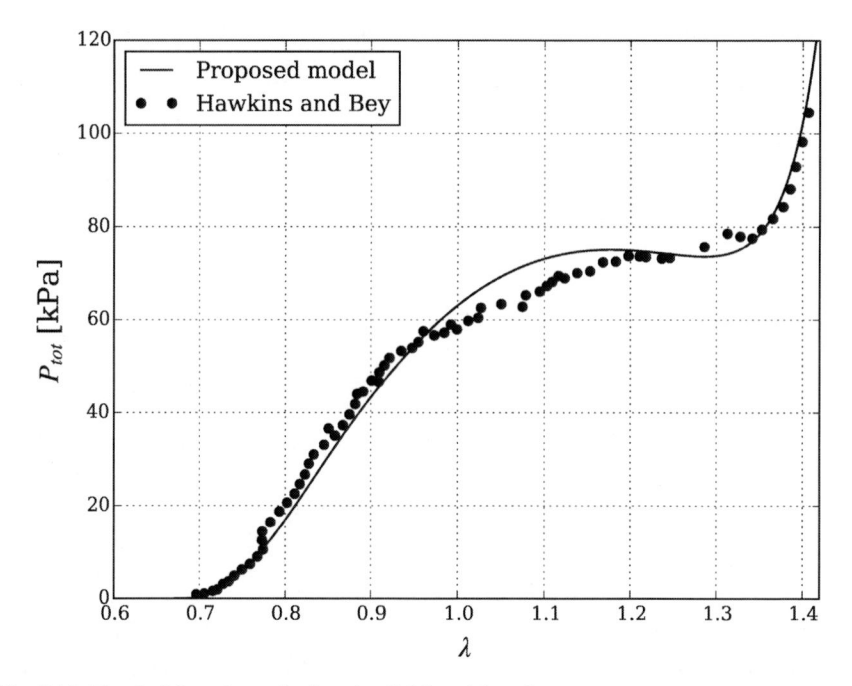

Fig. 7.11 Trend of P_{tot} when γ is given by (7.14) and λ varies

that the function defined in (7.14) is continuous; in particular we impose $\gamma(\lambda_{min}) = 0$, so that the starting value of activation does not change. Moreover, the function approximates very well the numerical values of γ in the range $0.7 < \lambda < 1.5$. However, the fitting is not so good when λ becomes larger: for instance, the function is negative for $\lambda \geq 1.6$. Nevertheless, the latter behavior of γ does not influence too much the curve P_{tot}, since in that region $P_{pas} \gg P_{act}$. Indeed, one can even neglect the activation for large stretches. The total stress response is plotted in Fig. 7.11 in comparison to the data given in [10].

Finally, it is interesting to run the simulations in the case of loss of activation, i.e. when the damage parameter d varies. In order to find the suitable activation function $\gamma(\lambda)$, it is sufficient to multiply the term P_{act} in (7.14) by $(1 - d)$. As one would expect from Fig. 7.9, we have that when d increases the activation γ decreases (Fig. 7.12a). This means that lowering the curve of P_{act} results in a decrease of $\gamma(\lambda)$, which leads to a lowered total stress response. As one can see in Fig. 7.12b, the damage parameter mainly affects the value of the stress in the region near λ_{opt}, where the active stress reaches its maximum. However, the qualitative behavior of the stress curve does not change, at least for $d \leq 0.5$. In particular, after a plateau, the stress follows the exponential growth of the passive curve.

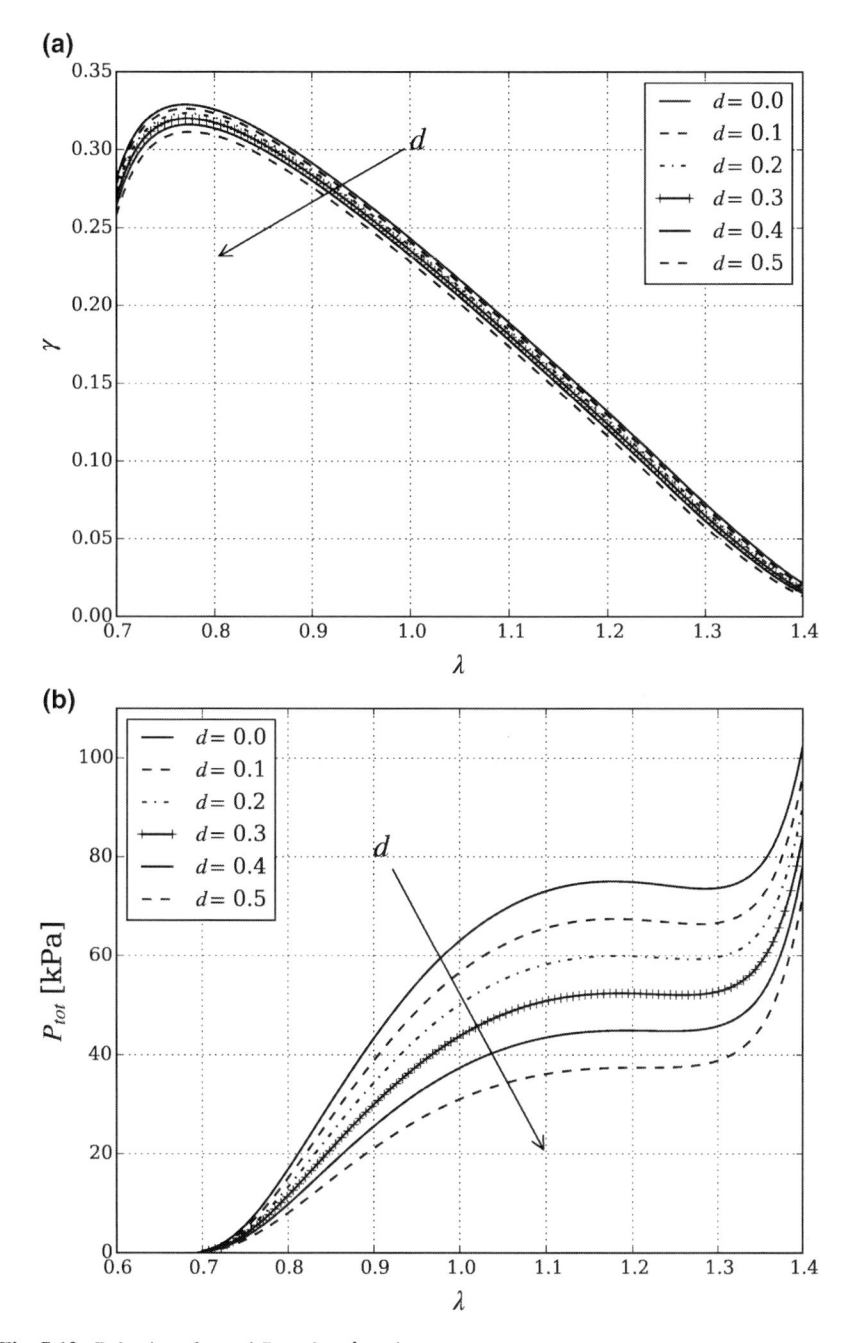

7 A Continuum Model of Skeletal Muscle Tissue with Loss of Activation

Fig. 7.12 Behavior of γ and P_{tot} when λ varies

Acknowledgements This work has been supported by the project *Active Ageing and Healthy Living* [19] of the Università Cattolica del Sacro Cuore and partially supported by GNFM (Gruppo Nazionale per la Fisica Matematica) of INdAM (Istituto Nazionale di Alta Matematica).
The authors wish to thank the anonymous referees for their useful comments.

References

1. Alnæs MS, Blechta J, Hake J, Johansson A, Kehlet B, Logg A, Richardson C, Ring J, Rognes ME, Wells GN (2015) The FEniCS project version 1.5. Arch Numer Softw 100:9–23
2. Ambrosi D, Pezzuto S (2012) Active stress vs. active strain in mechanobiology: constitutive issues. J Elast 107:199–212
3. Blemker SS, Pinsky PM, Delp SL (2005) A 3D model of muscle reveals the causes of nonuniform strains in the biceps brachii. J Biomech 38:657–665
4. Böl M, Reese S (2008) Micromechanical modelling of skeletal muscles based on the finite element method. Comput Methods Biomech Biomed Eng 11:489–504
5. Chagnon G, Rebouah M, Favier D (2015) Hyperelastic energy densities for soft biological tissues: a review. J Elast 120:129–160
6. Cruz-Jentoft AJ, Baeyens JP, Bauer JM, Boirie Y, Cederholm T, Landi F, Martin FC, Michel JP, Rolland Y, Schneider SM, Topinková E, Vandewoude M, Zamboni M (2010) Sarcopenia: European consensus on definition and diagnosis. Age Ageing 39:412–423
7. Ehret AE, Itskov M (2007) A polyconvex hyperelastic model for fiber-reinforced materials in application to soft tissues. J Mater Sci 42:8853–8863
8. Ehret AE, Itskov M (2009) Modeling of anisotropic softening phenomena: application to soft biological tissues. Int J Plast 25:901–919
9. Ehret AE, Böl M, Itskov M (2011) A continuum constitutive model for the active behaviour of skeletal muscle. J Mech Phys Solids 59:625–636
10. Hawkins D, Bey M (1994) A comprehensive approach for studying muscle-tendon mechanics. ASME J Biomech Eng 116:51–55
11. Heidlauf T, Röhrle O (2014) A multiscale chemo-electro-mechanical skeletal muscle model to analyze muscle contraction and force generation for different muscle fiber arrangements. Front Physiol 5:1–14
12. Hernández-Gascón B, Grasa J, Calvo B, Rodríguez JF (2013) A 3D electro-mechanical continuum model for simulating skeletal muscle contraction. J Theor Biol 335:108–118
13. Johansson T, Meier P, Blickhan R (2000) A finite-element model for the mechanical analysis of skeletal muscles. J Theor Biol 206:131–149
14. Lang T, Streeper T, Cawthon P, Baldwin K, Taaffe DR, Harris TB (2010) Sarcopenia: etiology, clinical consequences, intervention, and assessment. Osteoporos Int 21:543–559
15. Martins JAC, Pires EB, Salvado R, Dinis PB (1998) A numerical model of passive and active behavior of skeletal muscles. Comput Methods Appl Mech Eng 151:419–433
16. Musesti A, Giusteri GG, Marzocchi A (2014) Predicting ageing: on the mathematical modelization of ageing muscle tissue. In: Riva G et al. (eds.) Active ageing and healthy living, Chap 17. IOS press, Amsterdam
17. Nardinocchi P, Teresi L (2007) On the active response of soft living tissues. J Elast 88:27–39
18. Paetsch C, Trimmer BA, Dorfmann A (2012) A constitutive model for active-passive transition of muscle fibers. Int J Non-Linear Mech 47:377–387
19. Riva G, Ajmone Marsan P, Grassi C (2014) Active ageing and healthy living. IOS press, Amsterdam
20. Schröder J, Neff P (2003) Invariant formulation of hyperelastic transverse isotropy based on polyconvex free energy functions. Int J Solids Struct 40:401–445
21. Taber LA, Perucchio R (2000) Modeling heart development. J Elast 61:165–197

22. van Leeuwen JL (1991) Optimum power output and structural design of sarcomeres. J Theor Biol 149:229–256
23. van Leeuwen JL (1992) Muscle function in locomotion. In: Advances in comparative and environmental physiology, Chap. 7. Springer, Heidelberg
24. von Haehling S, Morley JE, Anker SD (2010) An overview of sarcopenia: facts and numbers on prevalence and clinical impact. J Cachexia Sarcopenia Muscle 1:129–133

Chapter 8
Optimal Control of Slender Microswimmers

Marta Zoppello, Antonio DeSimone, François Alouges, Laetitia Giraldi, and Pierre Martinon

8.1 Mathematical Setting of the Problem

In this section we describe the kinematics of the so called N-link swimmer, inspired by the Purcell's 3-link swimmer. A discrete representation of the swimmer's curvature is provided by the angles between successive links. These angles are considered as freely prescribed shape parameters. We then write the balance of total viscous force and torque, i.e. the equations of motion, solving for the time evolution of position and orientation of the swimmer in response to a prescribed history of (concentrated) curvatures along the swimmer's body.

M. Zoppello (✉)
Universitá degli studi di Padova, Via Trieste 63, 35121 Padova, Italy
e-mail: mzoppell@math.unipd.it

A. DeSimone
Scuola Internazionale di Studi Superiori Avanzati (SISSA), via Bonomea 265, 34136 Trieste, Italy
e-mail: desimone@sissa.it

F. Alouges
École Polytechnique CNRS, Route de Saclay, 91128 Palaiseau, France
e-mail: francois.alouges@polytechnique.edu

L. Giraldi
INRIA Sophia Antipolis Méditerranée, Team/équipe McTAO, B.P. 93, 06902 Sophia Antipolis, France
e-mail: laetitia.giraldi@inria.fr

P. Martinon
Team COMMANDS, INRIA Saclay - CMAP, École Polytechnique, Route de Saclay, 91128 Palaiseau, France
e-mail: pierre.martinon@inria.fr

8.1.1 Kinematics of the N-Link Swimmer

Here we are interested in essentially one-dimensional swimmers moving in a plane. This setting is suitable for the study of slender, one-dimensional swimmers exploring planar trajectories. The general case is a bit more involved because of the non-additivity of three-dimensional rotations, see e.g. [2], but it can be handled with similar techniques.

Our swimmer is composed of N rigid links with joints at their ends (see Fig. 8.1), moving in the plane $(\mathbf{e}_x, \mathbf{e}_y)$ (2d lab-frame). We set $\mathbf{e}_z := \mathbf{e}_x \times \mathbf{e}_y$. The i-th link is the segment with end points $\mathbf{x}_i$ and $\mathbf{x}_{i+1}$. It has length $L_i > 0$ and makes an angle θ_i with the vector $\mathbf{e}_x$. The length of the sticks is chosen such that the size of the swimmer is of order of micrometers (μm). We define by $\mathbf{x}_i := (x_i, y_i)$ ($i = 1, \cdots, N$) the coordinates of the first end of each link. Notice that, for $i \in \{2 \cdots N\}$, $\mathbf{x}_i$ is a function of $\mathbf{x}_1$, θ_k and L_k, with $k \in \{1 \cdots i-1\}$:

$$\mathbf{x}_i := \mathbf{x}_1 + \sum_{k=1}^{i-1} L_k \begin{pmatrix} \cos(\theta_k) \\ \sin(\theta_k) \end{pmatrix}. \tag{8.1}$$

The swimmer is described by two kind of variables:

- the state variables which denote the position and the orientation of one selected link, labeled as the $i*$-th one;
- the shape variables which describe the relative angles between successive links. For each link with $i > i*$, this is the angle relative to the preceding one, denoted by $\alpha_i = \theta_i - \theta_{i-1}$, for $i* < i \leq N$. For $i < i*$ this is the angle relative to the following one, denoted by $\alpha_i = \theta_{i+1} - \theta_i$, for $1 \leq i < i*$.

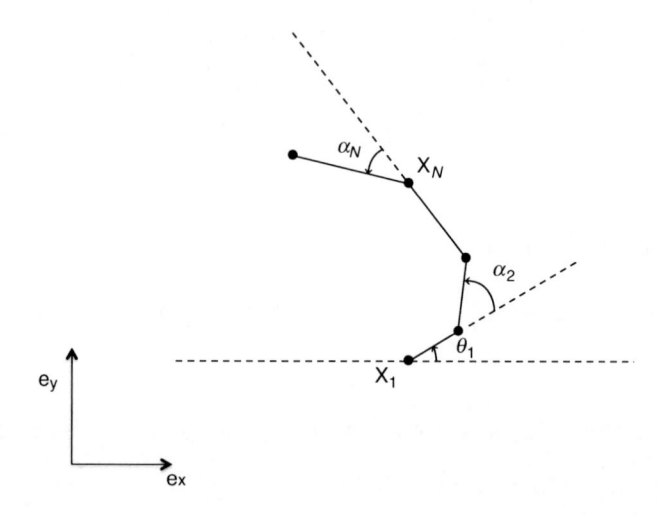

Fig. 8.1 Coordinates of the N-link swimmer. Reprinted from [1] with permission from Elsevier

For example, if the triplet (x_1, y_1, θ_1) is the state of the swimmer then the vector $(\alpha_2 = \theta_2 - \theta_1, \cdots, \alpha_N = \theta_N - \theta_{N-1})$ describes the shape of the swimmer. We will use these coordinates in the rest of the chapter.

8.1.2 Equations of Motion

The dynamics of the swimmer is governed by a system of three ODEs.

This system represents the Newton laws, in which inertia is neglected, namely

$$\begin{cases} \mathbf{F} = 0, \\ \mathbf{e}_z \cdot \mathbf{T}_{\mathbf{x}_1} = 0, \end{cases} \tag{8.2}$$

where $\mathbf{F}$ is the total force that the fluid exerts on the swimmer and $\mathbf{T}_{\mathbf{x}_1}$ is the corresponding total torque computed with respect to the point $\mathbf{x}_1$.

In what follows we use the local drag approximation of Resistive Force Theory (RTF), [9] to couple the fluid and the swimmer. This theory provides a simple and concise way to compute a local approximation of such forces, and it has been successfully used in several recent studies, see for example [4, 7]. According to this approximation the hydrodynamic forces are linear in the velocities of each point. More precisely denoting by s the arc length coordinate on the i-th link ($0 \le s \le L_i$), by $\mathbf{v}_i(s)$ the velocity of the corresponding point, and calling $\mathbf{e}_i = \begin{pmatrix} \cos(\theta_i) \\ \sin(\theta_i) \end{pmatrix}$ and $\mathbf{e}_i^\perp = \begin{pmatrix} -\sin(\theta_i) \\ \cos(\theta_i) \end{pmatrix}$ the unit vectors in the directions parallel and perpendicular to the i-th link respectively, we can write $\mathbf{x}_i(s) = \mathbf{x}_i + s\mathbf{e}_i$. Differentiating, we obtain,

$$\mathbf{v}_i(s) = \dot{\mathbf{x}}_i + s\dot{\theta}_i \mathbf{e}_i^\perp. \tag{8.3}$$

According to RTF, the density of the force $\mathbf{f}_i$ acting on the i-th segment depends linearly on the velocity and can be written as

$$\mathbf{f}_i(s) := -\xi \left(\mathbf{v}_i(s) \cdot \mathbf{e}_i \right) \mathbf{e}_i - \eta \left(\mathbf{v}_i(s) \cdot \mathbf{e}_i^\perp \right) \mathbf{e}_i^\perp, \tag{8.4}$$

where ξ and η are the drag coefficients in the directions of $\mathbf{e}_i$ and $\mathbf{e}_i^\perp$ respectively, measured in $\mathrm{N\,s\,m^{-2}}$ since it is homogeneous to mass divided by time^2. We thus obtain

$$\begin{cases} \mathbf{F} = \sum_{i=1}^{N} \int_0^{L_i} \mathbf{f}_i(s)\,ds, \\ \mathbf{e}_z \cdot \mathbf{T}_{\mathbf{x}_1} = \mathbf{e}_z \cdot \sum_{i=1}^{N} \int_0^{L_i} (\mathbf{x}_i(s) - \mathbf{x}_1) \times \mathbf{f}_i(s)\,ds. \end{cases} \tag{8.5}$$

Using (8.3) and (8.4) into (8.5), the total force and torque can be expressed as

$$\mathbf{F} = -\sum_{i=1}^{N} L_i \xi (\dot{\mathbf{x}}_i \cdot \mathbf{e}_i) \mathbf{e}_i + \left(L_i \eta (\dot{\mathbf{x}}_i \cdot \mathbf{e}_i^\perp) + \frac{L_i^2}{2} \eta \dot{\theta}_i \right) \mathbf{e}_i^\perp, \qquad (8.6)$$

and

$$\mathbf{e}_z \cdot \mathbf{T}_{\mathbf{x}_1} = -\sum_{i=1}^{N} \frac{L_i^2}{2} \eta \left(\dot{\mathbf{x}}_i \cdot \mathbf{e}_i^\perp \right) + \frac{L_i^3}{3} \eta \dot{\theta}_i$$

$$+ (\mathbf{x}_i - \mathbf{x}_1) \times \left(L_i \xi (\dot{\mathbf{x}}_i \cdot \mathbf{e}_i) \mathbf{e}_i + \left(L_i \eta (\dot{\mathbf{x}}_i \cdot \mathbf{e}_i^\perp) + \frac{L_i^2}{2} \eta \dot{\theta}_i \right) \mathbf{e}_i^\perp \right) \cdot \mathbf{e}_z. \qquad (8.7)$$

Moreover, the differentiation of (8.1) gives

$$\dot{\mathbf{x}}_i = \dot{\mathbf{x}}_1 + \sum_{k=1}^{i-1} L_k \dot{\theta}_k \mathbf{e}_k^\perp, \qquad (8.8)$$

which is linear in $\dot{\mathbf{x}}_1$ and $(\dot{\theta}_k)_{1 \leq k \leq N}$. This implies that also (8.6) and (8.7) are linear in $\dot{\mathbf{x}}_1$ and $\dot{\theta}_i$ for $i \in [1 \cdots N]$, and therefore system (8.2) reads

$$\begin{pmatrix} \mathbf{F} \\ \mathbf{e}_z \cdot \mathbf{T}_{\mathbf{x}_1} \end{pmatrix} = \mathbf{M}(\theta_1, \cdots, \theta_N) \begin{pmatrix} \dot{\mathbf{x}}_1 \\ \dot{\theta}_1 \\ \dot{\theta}_2 \\ \vdots \\ \dot{\theta}_N \end{pmatrix} = \begin{pmatrix} 0 \\ 0 \\ 0 \end{pmatrix}. \qquad (8.9)$$

We point out that for all $i \in \{2, \cdots, N\}$, $\alpha_i = \theta_i - \theta_{i-1}$, Eqs. (8.6) and (8.7) can be written using the relative angles $(\alpha_i)_{i=2,\cdots,N}$ instead of the variables $(\theta_i)_{2 \leq i \leq N}$ recalling that

$$\theta_i = \theta_{i-1} + \alpha_i, \quad i = 2, \cdots, N. \qquad (8.10)$$

To this end, we introduce the square matrix $\mathbf{C}$ defined by

$$\mathbf{C} = \begin{pmatrix} 1 & 0 & \cdots & \cdots & \cdots & \cdots & 0 \\ 0 & 1 & \ddots & & \ddots & \ddots & \vdots \\ 0 & 0 & 1 & \ddots & \ddots & & \vdots \\ 0 & 0 & -1 & \ddots & \ddots & & \vdots \\ \vdots & \vdots & 0 & \ddots & \ddots & & 0 \\ \vdots & \vdots & \vdots & \ddots & \ddots & \ddots & 0 \\ 0 & 0 & 0 & \cdots & 0 & -1 & 1 \end{pmatrix} \qquad (8.11)$$

and obtain

$$
\mathbf{C} \begin{pmatrix} \dot{\mathbf{x}}_1 \\ \dot{\theta}_1 \\ \dot{\theta}_2 \\ \vdots \\ \dot{\theta}_N \end{pmatrix} = \begin{pmatrix} \dot{\mathbf{x}}_1 \\ \dot{\theta}_1 \\ \dot{\alpha}_2 \\ \vdots \\ \dot{\alpha}_N \end{pmatrix}. \tag{8.12}
$$

Thus, by setting

$$\mathbf{N}(\theta_1, \alpha_2, \cdots, \alpha_N) := \mathbf{M}\left(\theta_1, \theta_2(\theta_1, \alpha_2, \cdots, \alpha_N), \cdots, \theta_N(\theta_1, \alpha_2, \cdots, \alpha_N)\right) \mathbf{C}^{-1}, \tag{8.13}$$

system (8.9) can be rewritten in the equivalent form

$$
\mathbf{N}(\theta_1, \alpha_2, \cdots, \alpha_N) \begin{pmatrix} \dot{\mathbf{x}}_1 \\ \dot{\theta}_1 \\ \dot{\alpha}_2 \\ \vdots \\ \dot{\alpha}_N \end{pmatrix} = \begin{pmatrix} 0 \\ 0 \\ 0 \end{pmatrix}. \tag{8.14}
$$

We observe that we can decompose the $3 \times (N+2)$ matrix $\mathbf{N}(\theta_1, \alpha_2, \cdots, \alpha_N)$ in blocks into a 3×3 sub-matrix $\mathbf{A}(\theta_1, \alpha_2, \cdots, \alpha_N)$ and a $3 \times (N-1)$ sub-matrix $\mathbf{B}(\theta_1, \alpha_2, \cdots, \alpha_N)$, according to

$$\mathbf{N} = (\mathbf{A} \,|\, \mathbf{B}) \,. \tag{8.15}$$

The matrix $\mathbf{A}$ is known as the "grand-resistance-matrix" of a rigid system evolving at frozen shape, i.e., with $\dot{\alpha}_i \equiv 0$, $i = 2, \ldots, N$, see [10]. It can be easily verified that it is symmetric and negative definite [10] and hence invertible. Therefore the equations of motion of the swimmer turn out to be affine system without drift. Indeed, solving (8.14) for $(\dot{\mathbf{x}}_1, \dot{\theta}_1)$ leads to

$$\begin{pmatrix} \dot{\mathbf{x}}_1 \\ \dot{\theta}_1 \end{pmatrix} = -\mathbf{A}^{-1}(\theta_1, \alpha_2, \cdots, \alpha_N)\, \mathbf{B}(\theta_1, \alpha_2, \cdots, \alpha_N) \begin{pmatrix} \dot{\alpha}_2 \\ \vdots \\ \dot{\alpha}_N \end{pmatrix}$$

which can be rewritten in the form

$$\begin{pmatrix} \dot{\mathbf{x}}_1 \\ \dot{\theta}_1 \end{pmatrix} = \sum_{i=2}^{N} \tilde{\mathbf{g}}_i(\theta_1, \alpha_2, \cdots, \alpha_N)\, \dot{\alpha}_i, \tag{8.16}$$

where the $N-1$ vector fields $\{\tilde{\mathbf{g}}_i\}_{i=2}^N$, are the columns of the $3 \times (N-1)$ matrix $-\mathbf{A}^{-1}\mathbf{B}$.

The equation above encodes the link between the displacement (both translation and rotation) of the swimmer and its deformation. More precisely, prescribing the shape functions $t \mapsto (\alpha_2, \cdots, \alpha_N)(t)$, the motion of the swimmer is obtained by solving the system (8.16). In what follows we call *stroke* a time-periodic shape change, i.e., the functions $t \mapsto \alpha_i(t)$, $i = 2, \cdots, N$ are all periodic, with the same period.

In order to solve (8.16) numerically, we need to compute the vector fields $\tilde{\mathbf{g}}_i$ explicitly. To this end, we notice that the total force $\mathbf{F}$ and the total torque $\mathbf{T}_{\mathbf{x}_1}$ depend linearly on $(\dot{\mathbf{x}}_i)_{1 \leq i \leq N}$ and $(\dot{\theta}_i)_{1 \leq i \leq N}$ and that these quantities depend in turn linearly on $(\dot{\mathbf{x}}_1, \dot{\theta}_1, \cdots, \dot{\theta}_N)$ in view of (8.8). Therefore, we can rewrite (8.6) and (8.7) as

$$\mathbf{F} = \mathbf{P}_1 \begin{pmatrix} \dot{\mathbf{x}}_1 \\ \vdots \\ \dot{\mathbf{x}}_N \\ -- \\ \dot{\theta}_1 \\ \vdots \\ \dot{\theta}_N \end{pmatrix} = \mathbf{P}_1 \mathbf{Q} \begin{pmatrix} \dot{\mathbf{x}}_1 \\ \dot{\theta}_1 \\ \vdots \\ \dot{\theta}_N \end{pmatrix}, \quad \mathbf{e}_z \cdot \mathbf{T}_{\mathbf{x}_1} = \mathbf{P}_2 \begin{pmatrix} \dot{\mathbf{x}}_1 \\ \vdots \\ \dot{\mathbf{x}}_N \\ -- \\ \dot{\theta}_1 \\ \vdots \\ \dot{\theta}_N \end{pmatrix} = \mathbf{P}_2 \mathbf{Q} \begin{pmatrix} \dot{\mathbf{x}}_1 \\ \dot{\theta}_1 \\ \vdots \\ \dot{\theta}_N \end{pmatrix},$$

(8.17)

where

$$\mathbf{P}_1 := \left(-\mathbf{m}_1 \cdots -\mathbf{m}_N \mid \tfrac{\eta}{2} L_1^2 \mathbf{e}_1^\perp \cdots \tfrac{\eta}{2} L_N^2 \mathbf{e}_N^\perp \right)$$

with $\mathbf{m}_i := L_i(\xi \mathbf{e}_i \otimes \mathbf{e}_i + \eta \mathbf{e}_i^\perp \otimes \mathbf{e}_i^\perp)$ for $i = 1 \cdots N$,

$$\mathbf{P}_2 := \left(\cdots -(L_i^2 \eta \mathbf{e}_i^\perp + (\mathbf{x}_i - \mathbf{x}_1) \times \mathbf{m}_i)^T \cdots \mid \cdots \eta L_i^2 (\tfrac{L_i}{3} + \tfrac{(\mathbf{x}_i - \mathbf{x}_1) \times \mathbf{e}_i^\perp \cdot \mathbf{e}_z}{2}) \cdots \right),$$

and, finally,

$$\mathbf{Q} = \begin{pmatrix} 1 & 0 & 0 & 0 & \cdots & 0 \\ 1 & L_1 \mathbf{e}_1^\perp & 0 & 0 & \cdots & 0 \\ 1 & L_1 \mathbf{e}_1^\perp & L_2 \mathbf{e}_2^\perp & 0 & \cdots & 0 \\ \vdots & \vdots & \vdots & \ddots & \cdots & 0 \\ 1 & L_1 \mathbf{e}_1^\perp & L_2 \mathbf{e}_2^\perp & \cdots & L_{N-1} \mathbf{e}_{N-1}^\perp & 0 \\ 0 & & & & & \\ \vdots & & & \text{Id} & & \\ 0 & & & & & \end{pmatrix}.$$

We thus have that the matrix **M** in (8.13) is

$$\mathbf{M} = \begin{pmatrix} \mathbf{P}_1 \mathbf{Q} \\ \mathbf{P}_2 \mathbf{Q} \end{pmatrix}$$

and we can compute $\mathbf{N} = \mathbf{C}^{-1}\mathbf{M}$, where $\mathbf{C}^{-1}$ is explicitly given as

$$\mathbf{C}^{-1} = \begin{pmatrix} 1 & 0 & \cdots & & & & 0 \\ 0 & 1 & \ddots & \ddots & \ddots & \ddots & \vdots \\ 0 & 0 & 1 & \ddots & \ddots & & \vdots \\ 0 & 0 & 1 & \ddots & \ddots & & \vdots \\ \vdots & \vdots & 1 & \ddots & \ddots & & 0 \\ \vdots & \vdots & \vdots & \ddots & \ddots & \ddots & 0 \\ 0 & 0 & 1 & \cdots & 1 & 1 & 1 \end{pmatrix}. \quad (8.18)$$

Matrices **A** and **B** are obtained from the columns of **N** as in (8.15) and, finally, the vectors $\tilde{\mathbf{g}}_i$ are simply the columns of $-\mathbf{A}^{-1}\mathbf{B}$. Finally the dynamics of the swimmer is expressed as

$$\begin{pmatrix} \dot{\alpha}_2 \\ \vdots \\ \dot{\alpha}_N \\ \dot{\mathbf{x}}_1 \\ \dot{\theta}_1 \end{pmatrix} = \sum_{i=1}^{N-1} \begin{pmatrix} \mathbf{b}_i \\ \tilde{\mathbf{g}}_i(\theta_1, \alpha_2, \cdots, \alpha_N) \end{pmatrix} \dot{\alpha}_{i+1}. \quad (8.19)$$

where $\mathbf{b}_i$ is the i-th vector of the canonical basis of $\mathbf{R}^{N-1}$.

8.2 Applications of the *N*-Link Swimmer

The *N*-link swimmer model is very useful and can be used as a discrete approximation of a swimmer's flexible tail whose shape is controlled by curvature. We show in this section, how curvature control can be implemented in our model in a concrete case reproducing the motion of a sperm cell analyzed in [7].

8.2.1 Curvature Approximation

Here, we describe the method to approximate the curvature of a beating tail with the discrete N link swimmer model. Let $L > 0$ be the total length of the flexible tail and let $\mathbf{r}(s, t)$ be the position, in the body frame of the swimmer (see Fig. 8.2), at time

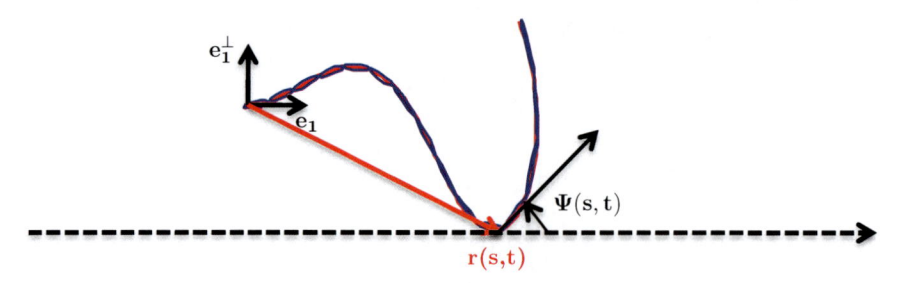

Fig. 8.2 The discrete approximation by the N-link swimmer (blue curve), $N = 15$ of a continuous tail (red curve). Reprinted from [1] with permission from Elsevier

$t > 0$ of the point of arc-length coordinate $s \in [0, L]$ along the tail. We also define the angle between the tangent vector to the tail at the point $\mathbf{r}(s,t)$ and the x-axis in the lab-frame as $\Psi(s,t)$. It is well known that the derivative of $\Psi(s,t)$ with respect to s is the local curvature of the curve.

We discretize the swimmer's tail into N equal parts of length $L_i = L/N$, and define the angles $(\theta_i)_{1 \leq i \leq N}$ by averaging the function $\Psi(s,t)$ on the interval $[(i-1)L/N, iL/N]$

$$\theta_i(t) = \frac{N}{L} \int_{\frac{(i-1)L}{N}}^{\frac{iL}{N}} \Psi(s,t)\, ds, \quad i = 1 \ldots N. \tag{8.20}$$

Finally, differentiating (8.20) with respect to time we get the angular velocities $\dot{\theta}_i$, $i = 1, \cdots, N$,

$$\dot{\theta}_i(t) = \frac{N}{L} \int_{\frac{(i-1)L}{N}}^{\frac{iL}{N}} \frac{\partial \Psi(s,t)}{\partial t}\, ds, \quad i = 1 \ldots N. \tag{8.21}$$

8.2.2 N-Link Approximation of Sperm Cell Swimmer

We now focus on reproducing with our model the motion of a sperm cell and compare to the one reported in [7]. To perform this comparison, we have to take into account the presence of the head of the sperm cell. In order to do this, we modify the first segment of the N-link swimmer so that it has its own translational and rotational viscous drag coefficients, $(\xi_{head}, \eta_{head})$ and ζ_{head} respectively. Indeed, we denote by $\mathbf{x}_1$ the position of the central point of the head and θ_1 the angle the direction of first segment (attached to the head) ($\mathbf{e}_1$) makes with the horizontal axis. The movement of the head generates a viscous force and torque that are given by

$$\mathbf{F}_{head} = -\xi_{head}(\dot{\mathbf{x}}_1 \cdot \mathbf{e}_1)\mathbf{e}_1 - \eta_{head}(\dot{\mathbf{x}}_1 \cdot \mathbf{e}_1^\perp)\mathbf{e}_1^\perp, \tag{8.22}$$

8 Optimal Control of Slender Microswimmers

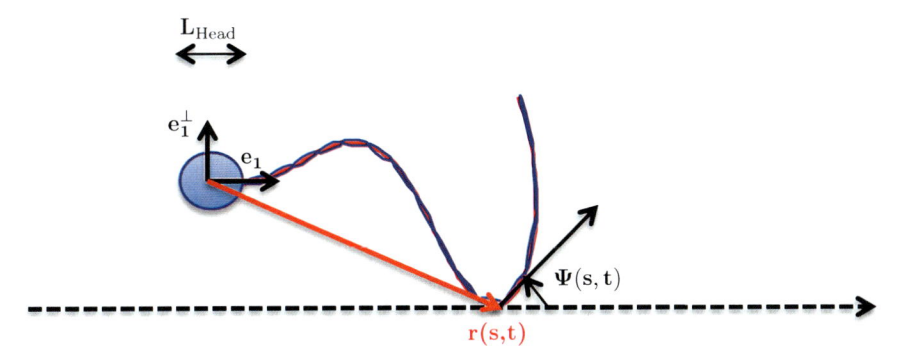

Fig. 8.3 The prescribed continuous wave (red curve) and its discrete approximation by the N-link swimmer (blue curve), $N = 15$. Reprinted from [1] with permission from Elsevier

and

$$\mathbf{T}_{head} \cdot \mathbf{e}_z = -\zeta_{head}\dot{\theta}_1 . \tag{8.23}$$

We also fix that the length of the head is $L_{head} = 10\,\mu\text{m}$ and we assume again that L is the length of the tail which is attached to one of the extremities of the head segment. As suggested in [13], the wave profile along the tail of the sperm cell swimmer was obtained from experimental data, keeping only the two first Fourier modes, and we use the method described before in Sect. 8.2.1 to approximate the tail's motion.

More precisely, we describe the shape of the wave shown in Fig. 8.3 by

$$\mathbf{r}(s,t) = \frac{L_{head}}{2}\mathbf{e}_1(t) + \int_0^s \cos(\Psi(u,t))\mathbf{e}_1(t) + \sin(\Psi(u,t))\mathbf{e}_1^\perp(t)du . \tag{8.24}$$

where

$$\Psi(s,t) = K_0 s + 2A_0 s \cos(\omega t - \frac{2\pi s}{\lambda}) . \tag{8.25}$$

In the previous equations, K_0 represents the mean flagellar curvature while ω, λ and A_0 are the frequency, the wave-length and the amplitude of the wave respectively. Following [7], in the next numerical simulations we use the following values for the wave parameters: $A_0 = 15.2 \cdot 10^3\,\text{rad}\,\text{m}^{-1}$, $K_0 = 19.1 \cdot 10^3\,\text{rad}\,\text{m}^{-1}$, $\omega = 200\,\text{rad}\,\text{s}^{-1}$ and $\lambda = 71.6 \cdot 10^{-6}\,\text{m}$.

Except the first segment, we discretize the rest of the tail with $N - 1$ segments of extremities $(\mathbf{x}_i, \mathbf{x}_{i+1})$ for $i = 2, \cdots, N$. We use the method described in Sect. 8.2.1, to approximate the beating wave and obtain the shapes shown in Fig. 8.4 for one period ($0 \leq t \leq \frac{2\pi}{\omega}$).

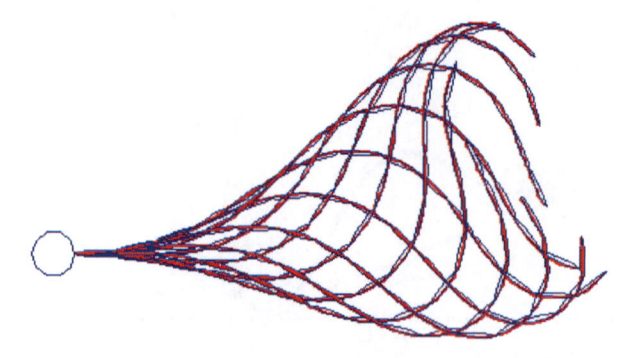

Fig. 8.4 Flagellar beating during one period. The red curve represents the tail as described by formula 8.24 while the blue links describe the tail according to our discrete approximation. Reprinted from [1] with permission from Elsevier

With the above notation, the equations of motion become

$$\begin{cases} \mathbf{F} = \mathbf{F}_{head} + \sum_{i=1}^{N} \int_0^{L_i} \mathbf{f}_i(s)\, ds\,, \\ \mathbf{T}_{\mathbf{x}_1} = \mathbf{T}_{head} + \sum_{i=1}^{N} \int_0^{L_i} \mathbf{f}_i(s) \times (\mathbf{x}_i(s) - \mathbf{x}_1)\, ds\,. \end{cases} \quad (8.26)$$

where $L_i = L/N$ is the length of each segment $(\mathbf{x}_i, \mathbf{x}_{i+1})$ for $i = 2, \cdots, N$, while the first segment, also of size $L_1 = L/N$ is given by $(\mathbf{x}_1 + \frac{L_{head}}{2}\mathbf{e}_1, \mathbf{x}_2)$.

Since the two previous formulas (8.26) are linear in $\dot{\theta}_1$ and $\dot{\mathbf{x}}_1$, we end up with the same compact expression of the equations of motion (8.16). More precisely, the matrix $\mathbf{P}_1$ and $\mathbf{P}_2$ defined in system (8.17) are replaced by

$$\mathbf{P}_1^{head} := \left(-\xi_{head}\mathbf{e}_1 \otimes \mathbf{e}_1 + \eta_{head}\mathbf{e}_1^\perp \otimes \mathbf{e}_1^\perp \; -\mathbf{m}_1 \; \cdots \; -\mathbf{m}_N \; | \; \tfrac{\eta}{2}L_1^2\mathbf{e}_1^\perp \; \cdots \; \tfrac{\eta}{2}L_N^2\mathbf{e}_N^\perp \right)$$

and

$$\mathbf{P}_2^{head} := \left(-p_1 \; \cdots \; -p_N \; | \; -\zeta_{head} + q_1 \; q_2 \; \cdots \; q_N \right),$$

with $\mathbf{m}_i := L_i(\xi \mathbf{e}_i \otimes \mathbf{e}_i + \eta \mathbf{e}_i^\perp \otimes \mathbf{e}_i^\perp)$ for $i = 1 \cdots N$, and $p_i := (L_i^2 \eta \mathbf{e}_i^\perp + (\mathbf{x}_i - \mathbf{x}_1) \times \mathbf{m}_i)^T$, $q_i := \eta L_i^2(\frac{L_i}{3} + \frac{(\mathbf{x}_i - \mathbf{x}_1) \times \mathbf{e}_i^\perp \cdot \mathbf{e}_z}{2})$, for $i = 1 \cdots N$.

Moreover we use the following values for the drag coefficients

- for the head, $\xi_{head} = 40.3 \cdot 10^3$ pN s m^{-1}, $\eta_{head} = 46.1 \cdot 10^3$ pN s m^{-1}, and $\zeta_{head} = 0.84 \cdot 10^{-6}$ pN s m
- for the links composing the tail, $\xi = 0.38 \cdot 10^9$ pN s m^{-2}, $\frac{\eta}{\xi} = 1.89$.

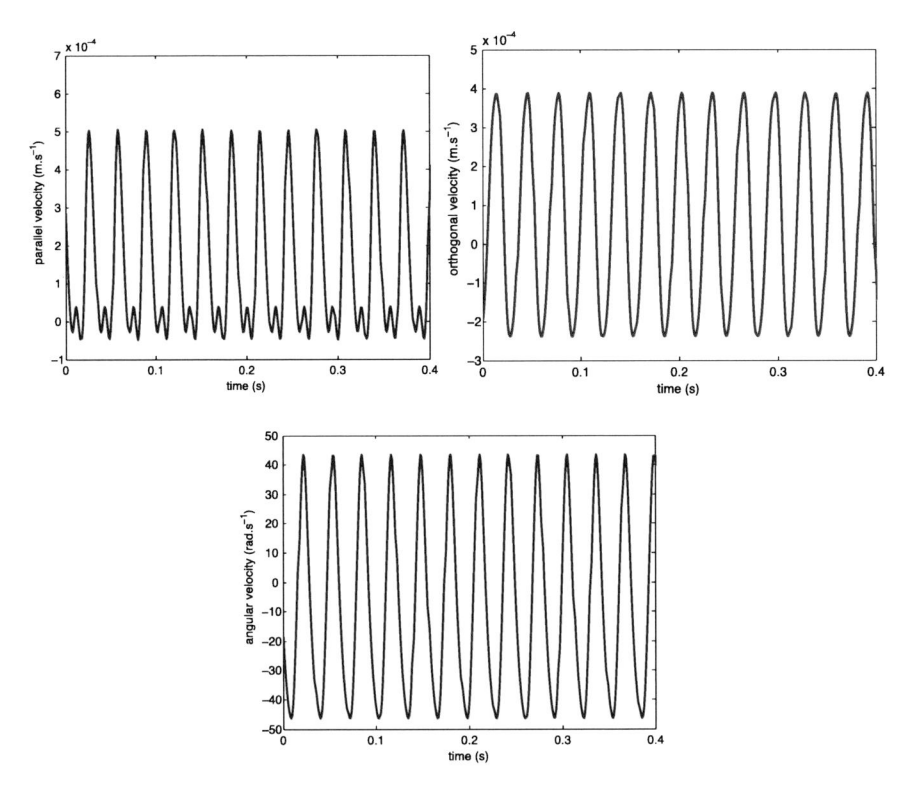

8 Optimal Control of Slender Microswimmers   171

Fig. 8.5 Above translational speed of the swimmer head in the tangent and perpendicular directions, and below rotational speed. Reprinted from [1] with permission from Elsevier

The graphs in Figs. 8.5 and 8.6 below, summarize our results that are in perfect agreement with those of [7] (see Figure 3 for the trajectory and Figure 4 for the various speeds).

8.3 Controllability

This section is devoted to the controllability of the N-link swimmer, which is its ability to move between two fixed configurations prescribing (controlling) its shape parameters. More precisely we prove that there exist control shape functions which allow the swimmer to move everywhere in the plane.

8.3.1 Classical Results in Geometric Control

We start by recalling in this subsection some classical results in geometric control theory which we will use later.

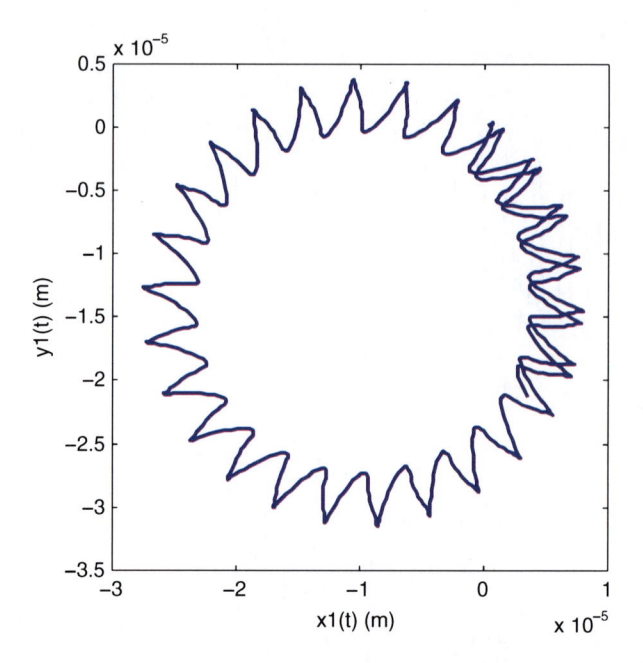

Fig. 8.6 Trajectory of the head of the sperm-cell during one period. Reprinted from [1] with permission from Elsevier

Theorem 8.1 (Chow (See [6])) *Let* $m,n \in \mathbb{N}$ *and let* $(\mathbf{g}_i)_{i=1,n}$ *be* $\mathscr{C}^\infty$ *vector fields on* $\mathbb{R}^n$. *Consider the control system, of state trajectory* $\mathbf{q}$,

$$\dot{\mathbf{q}} = \sum_{i=1}^{m} u_i \mathbf{g}_i(\mathbf{q}), \qquad (8.27)$$

with input function $\mathbf{u} = (u_i)_{i=1,m} \in L^\infty\left([0, +\infty[, \mathbf{B}_{\mathbb{R}^n}(0,\delta)\right)$ *for some* $\delta > 0$.

Let $\mathscr{O}$ *an open and connected set of* $\mathbb{R}^n$ *and assume that*

$$\mathbf{Lie_q}(\mathbf{g}_1, \ldots \mathbf{g}_m) = \mathbb{R}^n \quad \mathbf{q} \in \mathscr{O}$$

Then the system (8.27) is controllable, i.e., for every $\mathbf{q}_0$, $\mathbf{q}_1$ *in* $\mathscr{O}$ *and for every* $T > 0$ *exists* $\mathbf{u} \in L^\infty((0,T), \mathbf{B}_{\mathbb{R}^n}(0,\delta))$ *such that* $\mathbf{q}(0) = \mathbf{q}_0$ *and* $\mathbf{q}(T) = \mathbf{q}_1$ *and* $\mathbf{q}(t) \in \mathscr{O}$ *for every* $t \in [0,T]$.

If the vector fields are analytic, we can apply the Orbit Theorem to extend the dimension property of the Lie algebra defined by the dynamics vector fields on the whole orbit.

Theorem 8.2 (Orbit (See [11])) *Let* $\mathscr{M}$ *be an analytic manifold, and* $\mathscr{F}$ *a family of analytic vector fields on* $\mathscr{M}$. *Then*

a) each orbit of $\mathscr{F}$ is an analytic submanifold of $\mathscr{M}$, and
b) if N is an orbit of $\mathscr{F}$, then the tangent space of N at x is given by $Lie_x(\mathscr{F})$. In particular the dimension of $Lie_x(\mathscr{F})$ is constant as x varies on N.

8.3.2 Main Theorem

Theorem 8.3 *Consider the N-link swimmer described in Sect. 8.1 evolving in the space $\mathbf{R}^2$. Then for almost every lengths of the sticks $(L_i)_{i=1,\cdots,N}$ and for any initial configuration $(\mathbf{x}_1^i, \theta_1^i, \alpha_2^i, \cdots, \alpha_N^i) \in \mathbf{R}^2 \times [0, 2\pi]^N$, any final configuration $(\mathbf{x}_1^f, \theta_1^f, \alpha_2^f, \cdots, \alpha_N^f)$ and any final time $T > 0$, there exists a shape function $(\alpha_2, \cdots, \alpha_N) \in \mathscr{W}^{1,\infty}([0, T])$, satisfying $(\alpha_2, \cdots, \alpha_N)(0) = (\alpha_2^i, \cdots, \alpha_N^i)$ and $(\alpha_2, \cdots, \alpha_N)(T) = (\alpha_2^f, \cdots, \alpha_N^f)$ and such that if the self-propelled swimmer starts in position $(\mathbf{x}_1^i, \theta_1^i)$ with the shape $(\alpha_2^i, \cdots, \alpha_N^i)$ at time $t = 0$, it ends at position $(\mathbf{x}_1^f, \theta_1^f)$ and shape $(\alpha_2^f, \cdots, \alpha_N^f)$ at time $t = T$ by changing its shape along $(\alpha_2, \cdots, \alpha_N)(t)$.*

Proof The proof of the theorem is divided into three steps. First of all, we show the analyticity of the dynamics vector fields. Then, we prove the controllability of the 3-link swimmer (Purcell swimmer), exploiting the Chow theorem and the Orbit theorem. Finally, we generalize the result to the case of N links.

In our case, the manifold in which the state and the shape of the swimmer evolve is defined by $\mathscr{M} := [0, 2\pi]^{N-1} \times \mathbb{R}^2 \times [0, 2\pi]$. The vector fields of the dynamics are denoted by

$$\mathbf{g}_i(\theta_1, \alpha_2, \cdots, \alpha_N) := \begin{pmatrix} \mathbf{b}_i \\ \tilde{\mathbf{g}}_i(\theta_1, \alpha_2, \cdots, \alpha_N) \end{pmatrix}.$$

We say that the Lie algebra of the family of vector fields $\{\mathbf{g}_i\}_{i=1,\cdots,N-1}$ is fully generated at the point $\mathbf{q} = (\alpha_2, \cdots, \alpha_N, x_1, y_1, \theta_1) \in \mathscr{M}$ if the tangent space of the manifold, $T_{\mathbf{q}}\mathscr{M}$, is equal to the Lie algebra $Lie((\mathbf{g}_i))_{i=1,\cdots,N-1}(\mathbf{q})$.

8.3.2.1 Regularity

We first prove that the vector fields $(\tilde{\mathbf{g}}_i)$ are analytic on $\mathscr{M}$. From (8.6) and (8.7), the entries of the matrices $\mathbf{A}$ and $\mathbf{B}$ are analytic functions on $[0, 2\pi]^N$. Since the coefficients of $\mathbf{A}^{-1}$ are obtained by multiplication and division of those of $\mathbf{A}$, and because $det(\mathbf{A}) \neq 0$ ($\mathbf{A}$ is symmetric and negative defined), the entries of $\mathbf{A}^{-1}$ remain analytic functions on $[0, 2\pi]^N$. Thus, the $(\tilde{\mathbf{g}}_i)_{i=1,\cdots,N} := \mathbf{A}^{-1}\mathbf{B}$ are analytic on $[0, 2\pi]^N$.

8.3.2.2 Controllability of the Purcell Swimmer ($N = 3$)

Setting $N = 3$ in (8.19) the dynamics becomes

$$\begin{pmatrix} \dot{\alpha}_2 \\ \dot{\alpha}_3 \\ \dot{x}_1 \\ \dot{y}_1 \\ \dot{\theta}_1 \end{pmatrix} = \mathbf{g}_1(\theta_1, \alpha_2, \alpha_3)\dot{\alpha}_2 + \mathbf{g}_2(\theta_1, \alpha_2, \alpha_3)\dot{\alpha}_3 . \tag{8.28}$$

To prove the controllability of this system we want use Theorem 8.1.

Therefore we compute the Lie algebra of the vector fields $\mathbf{g}_1$ and $\mathbf{g}_2$ for any $\theta_1 \in [0, 2\pi]$ at $(\alpha_2, \alpha_3) = (0, 0)$, for a swimmer whose sticks have the length $L_1 = L_3 = L$ and $L_2 = 2L$ where $L > 0$. First we have

$$\mathbf{g}_1(\theta_1, 0, 0) = \begin{pmatrix} 1 \\ 0 \\ \frac{9L\sin(\theta_1)}{64} \\ -\frac{9L\cos(\theta_1)}{64} \\ \frac{27}{32} \end{pmatrix} \quad \mathbf{g}_2(\theta_1, 0, 0) = \begin{pmatrix} 0 \\ 1 \\ -\frac{7L\sin(\theta_1)}{64} \\ \frac{7L\cos(\theta_1)}{64} \\ -\frac{5}{32} \end{pmatrix}$$

Then, the iterated Lie brackets are equals to

$$[\mathbf{g}_1, \mathbf{g}_2](\theta_1, 0, 0) = \left(0, 0, \frac{7L(\eta-\xi)\cos(\theta_1)}{128\xi}, \frac{7L(\eta-\xi)\sin(\theta_1)}{128\xi}, 0\right)^T ,$$

$$[\mathbf{g}_1, [\mathbf{g}_1, \mathbf{g}_2]](\theta_1, 0, 0) = \begin{pmatrix} 0 \\ 0 \\ -\frac{L(126\eta^2+31\xi\eta-76\xi^2)\sin(\theta_1)}{4096\eta\xi} \\ \frac{L(126\eta^2+31\xi\eta-76\xi^2)\cos(\theta_1)}{4096\eta\xi} \\ -\frac{3(9\eta^2-4\xi\eta+4\xi^2)}{2048\eta\xi} \end{pmatrix} ,$$

$$[\mathbf{g}_2, [\mathbf{g}_1, \mathbf{g}_2]](\theta_1, 0, 0) = \begin{pmatrix} 0 \\ 0 \\ \frac{L(36\eta^2-103\xi\eta+148\xi^2)\sin(\theta_1)}{4096\eta\xi} \\ -\frac{L(36\eta^2-103\xi\eta+148\xi^2)\cos(\theta_1)}{4096\eta\xi} \\ \frac{3(9\eta^2-4\xi\eta+4\xi^2)}{2048\eta\xi} \end{pmatrix} .$$

To see if they are linearly independent we compute the determinant of the matrix whose columns are the five previous vector fields that is equal to

$$\frac{21L^2(\eta-\xi)^2(45\eta+112\xi)\left(9\eta^2-4\eta\xi+4\xi^2\right)}{536870912\eta^2\xi^3}. \tag{8.29}$$

Since the drag coefficients ξ and η are positive, this determinant is null only when $\xi = \eta$. This would imply an isotropic drag, as we would have if we use spheres instead of sticks. Thus in our case the Lie algebra of the vector fields $\mathbf{g}_1$ and $\mathbf{g}_2$ is fully generated at the point $(\theta_1, 0, 0)$, for any $\theta_1 \in [0, 2\pi]$.

Notice that any point $(\alpha_2, \alpha_3, \mathbf{x}_1, \theta_1) \in [0, 2\pi]^2 \times \mathbf{R}^2 \times [0, 2\pi]$ belongs to the orbit of the point $(0, 0, \mathbf{x}_1, \theta_1)$. Since the vector fields are analytic, the Orbit Theorem 8.2 guarantees that the Lie algebra of $\mathbf{g}_1$ and $\mathbf{g}_2$ is fully generated everywhere in the manifold $\mathcal{M} = [0, 2\pi]^2 \times \mathbf{R}^2 \times [0, 2\pi]$.

To conclude, by Chow Theorem (8.1) we get the controllability of the Purcell swimmer.

8.3.2.3 Controllability of the N-Link Swimmer

The third step is to generalize the previous controllability result to the N-link swimmer, whose dynamics is described by (8.19). It is easy to see that the vector fields $\mathbf{g}_i$ generate the tangent space of the manifolds $[0, 2\pi]^{N-1}$,

$$\text{Span}(\mathbf{g}_1, \cdots, \mathbf{g}_{N-1}) = \mathbf{R}^{N-1}. \tag{8.30}$$

We can obtain the vector fields $\mathbf{g}_1$ and $\mathbf{g}_2$ starting from the Purcell's one defined in (8.28) as follows: we add $N-2$ rows of zeroes, take sticks of null length $L_i = 0$ for $4 \leq i \leq N-1$, while keeping the three sticks $L_1 = L_3 = L$ and $L_2 = 2L$.

In this case, for any $(\mathbf{x}_1, \theta_1) \in \mathbf{R}^2 \times [0, 2\pi]$, Sect. 8.3.2.2 shows that $\mathbf{g}_1(\theta_1, 0, \cdots 0)$, $\mathbf{g}_2(\theta_1, 0, \cdots 0)$ and their iterated Lie brackets $[\mathbf{g}_1, \mathbf{g}_2](\theta_1, 0, \cdots 0)$, $[\mathbf{g}_1, [\mathbf{g}_1, \mathbf{g}_2]](\theta_1, 0, \cdots 0)$, and $[\mathbf{g}_2, [\mathbf{g}_1, \mathbf{g}_2]](\theta_1, 0, \cdots 0)$ are linearly independent.

Therefore, the Lie algebra of the family $(\mathbf{g}_i)_{i=1,\cdots,N-1}$ at the point $(\theta_1, 0, \cdots, 0)$ is equal to the tangent space $T_{(0,\cdots,0,\mathbf{x}_1,\theta_1)}\mathcal{M}$. Then, by analyticity of the vector fields $\mathbf{g}_i$, the Orbit Theorem ensures that the Lie algebra is fully generated everywhere for a swimmer whose length of sticks verify $L_1 = L_3 = L$, $L_2 = 2L$ and $L_{i \geq 4} = 0$.

We call $D^{(0,\cdots,0)}$, the function that maps $(L_1, \cdots, L_N)$ to the determinant of the vectors $\mathbf{g}_1, \cdots, \mathbf{g}_{N-1}$ and their iterated Lie brackets at the point $(0, \cdots, 0)$. Since the vector fields $\mathbf{g}_i$ depend analytically on the sticks length L_i, we get the analyticity of the function $D^{(0,\cdots,0)}$. Thus for any $L > 0$, the value of $D^{(0,\cdots,0)}$ at the point $(L, 2L, L, 0 \cdots 0)$ is not null. By analyticity, it remains non null almost everywhere in $\mathbf{R}^N$. Therefore, we obtain that the Lie algebra has full rank for almost every swimmer.

Finally, Chow Theorem gives the controllability in the Theorem 8.3. □

8.4 Minimum Time Optimal Control Problem for the *N*-Link Swimmer

We present in Sect. 8.4.1 the minimum time optimal control problem for the *N*-link swimmer, which is well defined from the controllability result proven in Sect. 8.3. Then in Sect. 8.4.2 we present the numerical method used to solve this problem.

8.4.1 Minimum Time Problem

For any time $t > 0$, we use the following notation: the state of the swimmer is $\mathbf{z}(t) := (\alpha_2, \cdots, \alpha_N, \mathbf{x}_1, \theta_1)(t)$, the control functions are $\mathbf{u}(t) := (\dot{\alpha}_2, \cdots, \dot{\alpha}_N)(t)$ and the dynamics is $\mathbf{f}(\mathbf{z}(t), \mathbf{u}(t)) = \sum_{i=1}^{N-1} \mathbf{g}_i(\mathbf{z}(t)) \dot{\alpha}_{i+1}(t)$.

We now assume that the swimmer starts at the initial configuration $\mathbf{z}^i$, and we fix a final state $\mathbf{z}^f$. Our aim is to find a swimming strategy that minimizes the time to swim between the initial and the final configuration, i.e.,

$$(OCP) \begin{cases} \inf t_f, \\ \dot{\mathbf{z}}(t) = \mathbf{f}(\mathbf{z}(t), \mathbf{u}(t)), \ \forall t \in [0, t_f], \\ \mathbf{u}(t) \in \mathbf{U} := [-1, 1]^N, \ \forall t \in [0, t_f], \\ \mathbf{z}(0) = \mathbf{z}^i, \quad \mathbf{z}(t_f) = \mathbf{z}^f. \end{cases}$$

By applying Filippov-Cesary Theorem [14] which ensures the existence of a solution of the minimum time problem for controllable systems, there exists a minimal time such that the constraints are satisfied i.e., the infimum can be written as a minimum.

8.4.2 Numerical Optimization

In order to solve this optimal control problem, we use a direct approach. This approach transforms the infinite dimensional optimal control problem (*OCP*) into a finite dimensional optimization problem (*NLP*). This is done with a discretization procedure on the dynamics equation summarized below:

$$\begin{aligned}
t \in [0, tf] &\to \{t_0 = 0, \ldots, t_N = tf\} \\
z(\cdot), u(\cdot) &\to X = \{z_0, \ldots, z_N, u_0, \ldots, u_{N-1}, tf\} \\
\text{Criterion} &\to \min tf \\
\text{Dynamics} &\to (ex: Euler) \ z_{i+i} = z_i + hf(z_i, u_i) \\
\text{Controls} &\to -1 \leq u_i \leq 1 \\
I/F \ \text{Cond.} &\to \Phi(z_0, z_N) = 0
\end{aligned}$$

8 Optimal Control of Slender Microswimmers

We therefore obtain a nonlinear programming problem on the discretized state and control variables

$$(NLP) \begin{cases} \min \ F(z) = tf \\ LB \leq C(z) \leq UB \end{cases}$$

All tests were run using the software BOCOP [5]. The discretized nonlinear optimization problem is solved by the well-known solver IPOPT [15] with MUMPS [3], while the derivatives are computed by sparse automatic differentiation with ADOL-C [16] and COLPACK [8]. In the numerical experiments, we used a Midpoint (implicit 2nd order) discretization with 1000 time steps. Execution times on a Xeon 3.2 GHz CPU were a few minutes.

8.5 Numerical Simulations for the Purcell's Swimmer (N = 3)

We present in this section the numerical simulations regarding the Purcell swimmer (three links). Without making any assumptions on the structure of the optimal trajectory, we obtain an optimal solution with periodic strokes. Comparing this stroke to the one of Purcell [4, 12], we observe that it gives a better displacement speed.

In the rest of the chapter, in order to match the notations used in [4], we will use the following coordinates (see Fig. 8.7):

- the position (x_2, y_2) of the center of the second stick, and its angle with the x-axis $\theta_2 := \theta_1 - \alpha_2$.
- the shape angles $\beta_1 := -\alpha_2$ and $\beta_3 := \alpha_3$.

Fig. 8.7 Purcell's 3-link swimmer

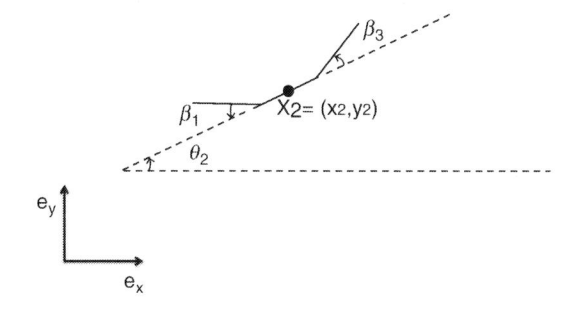

This reformulation gives the new dynamics

$$\begin{pmatrix} \dot{\beta}_1 \\ \dot{\beta}_3 \\ \dot{\mathbf{x}}_2 \\ \dot{\theta}_2 \end{pmatrix} = \mathbf{M}(\theta_2, \beta_1) \begin{pmatrix} \dot{\alpha}_2 \\ \dot{\alpha}_3 \\ \dot{\mathbf{x}}_1 \\ \dot{\theta}_1 \end{pmatrix},$$

$$\mathbf{M}(\theta_2, \beta_1) = \begin{pmatrix} -1 & 0 & 0 & 0 & 0 \\ 0 & 1 & 0 & 0 & 0 \\ \sin(\theta_2) + \cos(\beta_1) & 0 & 1 & 0 & -\sin(\theta_2) \\ -\cos(\beta_1) - \cos(\theta_2) & 0 & 0 & 1 & \cos(\theta_2) \\ -1 & 0 & 0 & 0 & 1 \end{pmatrix}.$$

As a result, the dynamics (8.19) reads in this case as

$$\begin{pmatrix} \dot{\beta}_1 \\ \dot{\beta}_3 \\ \dot{\mathbf{x}}_2 \\ \dot{\theta}_2 \end{pmatrix} = \tilde{\mathbf{f}}_1(\theta_2, \beta_2, \beta_3) \, \dot{\beta}_1 + \tilde{\mathbf{f}}_2(\theta_2, \beta_2, \beta_3) \, \dot{\beta}_3 \qquad (8.31)$$

where for $i = 1, 2$

$$\tilde{\mathbf{f}}_i(\theta_2, \beta_1, \beta_3) = \mathbf{M}(\theta_2, \beta_1,) \, \tilde{\mathbf{g}}_i(\theta_1, \alpha_2, \alpha_3). \qquad (8.32)$$

Observe that since the new state variables are the image of the former ones through a one-to-one mapping, the controllability result in Sect. 8.3.2.2 holds also for (8.31).

8.5.1 The Classical Purcell Stroke

We recall the stroke presented by Purcell in [12] in order to compare it to the optimal strategy given by our numerical results. Let us denote by $\Delta\theta$ the angular excursion, which means that β_1 and β_3 belong to $[-\frac{\Delta\theta}{2}, \frac{\Delta\theta}{2}]$. The Purcell stroke is defined by this periodic of deformation over $[0, T]$:

$$(\beta_1(t), \beta_3(t)) = \begin{cases} (\frac{4\Delta\theta}{T}t - \frac{\Delta\theta}{2}, \frac{\Delta\theta}{2}) & 0 \leq t \leq \frac{T}{4} \\ (\frac{\Delta\theta}{2}, -\frac{4\Delta\theta}{T}t + \frac{3\Delta\theta}{2}) & \frac{T}{4} \leq t \leq \frac{T}{2} \\ (-\frac{4\Delta\theta}{T}t + \frac{5\Delta\theta}{2}, -\frac{\Delta\theta}{2}) & \frac{T}{2} \leq t \leq \frac{3T}{4} \\ (-\frac{\Delta\theta}{2}, \frac{4\Delta\theta}{T}t - \frac{7\Delta\theta}{2}) & \frac{3T}{4} \leq t \leq T \end{cases}.$$

In what follows, we call the "classical" Purcell stroke the one corresponding to $\Delta\theta = \frac{\pi}{3}$, with $T = 4\Delta\theta$ chosen in order to satisfy the constraints on the controls of (OCP), i.e., $u_i(t) := \dot{\beta}_i(t) \in [-1, 1]$.

8.5.2 Comparison of the Optimal Stroke and Purcell Stroke

We set the initial position $(\mathbf{x}_2, \theta_2) = (0,0,0)$ and the final position $(\mathbf{x}_2, \theta_2) = (-0.25, 0, 0)$. Moreover we constrain the angles β_1 and β_3 in $[-\frac{\pi}{6}, \frac{\pi}{6}]$ for all time. Solving the minimum time problem numerically with the direct method gives us a solution that is actually periodic, as shown on Fig. 8.8. We notice that the x-displacement is not monotone: during each stroke, the swimmer alternately moves forward, closer to the target, and goes partially backward.

Now we isolate only one stroke from this solution, and compare it with the Purcell stroke. We show on Fig. 8.9 the angles functions β_1 and β_3, as well as the phase portrait. Note that the time required to complete our candidate for an optimal

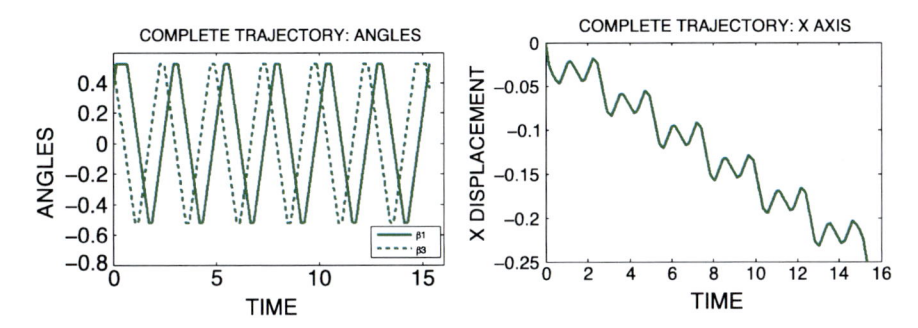

Fig. 8.8 Angles and x-displacement for a whole periodic trajectory

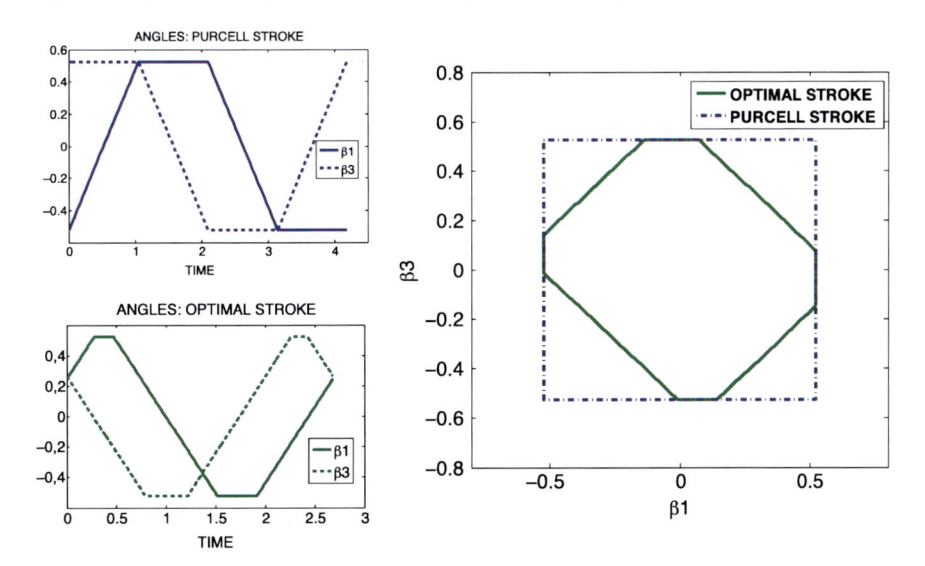

Fig. 8.9 Angles and phase portrait—Purcell stroke and optimal stroke

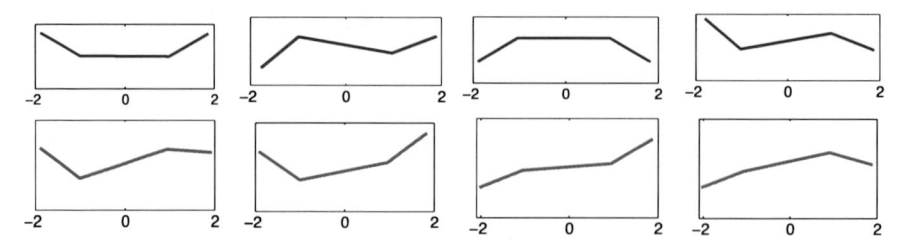

Fig. 8.10 Purcell stroke (above) and optimal stroke (below)

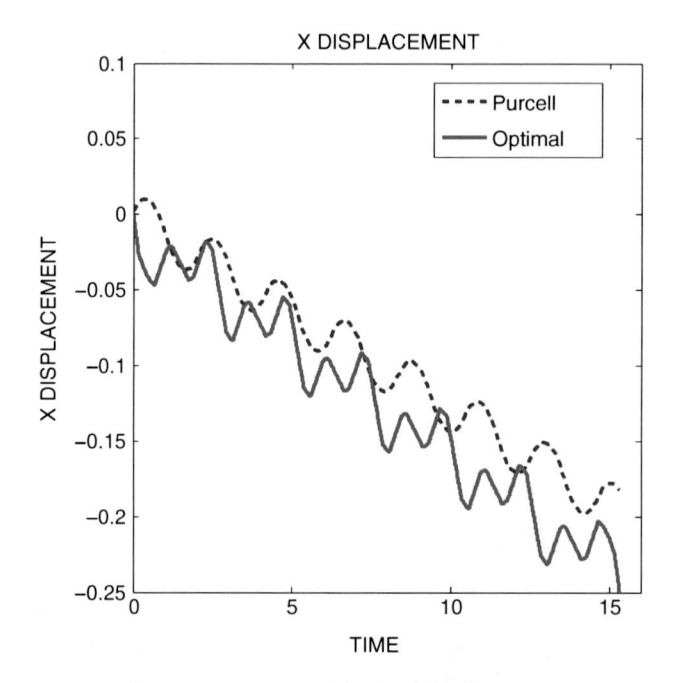

Fig. 8.11 x displacement for one Purcell and one optimal stroke

stroke is shorter than for the Purcell one (roughly 2.5 versus 4.1). We illustrate on Fig. 8.10 the shape changes in the plane for the Purcell and optimal stroke.

Finally we make a comparison between the two x-displacement, Fig. 8.11 shows the x-displacement of the swimmer with the classical Purcell stroke (dashed) and the optimal stroke (solid). Both trajectories were recomputed in Matlab using the same ODE solver, and the results for the Purcell stroke match the ones in [4]. The final time $t_f = 15.3252$ is the one given by the numeric simulation to reach $\mathbf{x}_2 = (-0.25, 0)$. We see that using Purcell strokes, the swimmer only reaches ($\approx -0.18, 0$), which confirms that our optimal stroke allows a greater x-displacement.

More precisely, each optimal stroke gives a x-displacement close to the Purcell stroke, however the cycle of deformation is performed in less time. Therefore, for a given time frame, more optimal strokes can be performed, leading to an overall greater displacement. In Fig. 8.11, almost 3.5 Purcell strokes are performed, while 6 optimal strokes are completed within the same time.

Remark The initial shape of the swimmer is not identical for both strategies, however the increasing gap between the two curves clearly shows that the optimal stroke is faster.

We also observe that the optimal stroke consistently gives a swimming speed better by 20% than the Purcell stroke.

8.6 Conclusions

In this chapter we study the N-link swimmer, and use the Resistive Force Theory to derive its dynamics. In this context, we prove that for N greater than 3 and for almost any N-uplet of sticks lengths, the swimmer is globally controllable in the whole plane. Then, we focus on finding a swimming strategy that leads the N-link swimmer from an fixed initial position to a given final position, in minimum time. As a consequence of the controllability result, we show that there exists a shape change function which allows to reach the final state in a minimal time. We formulate this optimal control problem and solve it with a direct approach (BOCOP) for the case $N = 3$ (Purcell swimmer). Without any assumption on the structure of the trajectory, we obtain a periodic solution, from which we identify an optimal stroke. Comparing this optimal stroke with the Purcell one confirms that it is better, actually giving a greater displacement speed.

References

1. Alouges F, DeSimone A, Giraldi L, Zoppello M (2013) Self-propulsion of slender microswimmers by curvature control: N-link swimmers. Int J Non Linear Mech 56(Supplement C):132–141. https://doi.org/10.1016/j.ijnonlinmec.2013.04.012
2. Alouges F, DeSimone A, Heltai L, Lefebvre A, Merlet B (2013) Optimally swimming Stokesian robots. Discret Contin Dyn Syst B 18:1189–1215
3. Amestoy P, Duff I, Koster J, L'Excellent J (2001) A fully asynchronous multifrontal solver using distributed dynamic scheduling. SIAM J Matrix Anal Appl 23(1):15–41
4. Becker L, Koehler S, Stone H (2003) On self-propulsion of micro-machines at low Reynolds number: Purcell's three-link swimmer. J Fluid Mech 490:15–35
5. Bonnans F, Martinon P, Grélard V (2012) Bocop - a collection of examples. Technical Report RR-8053, INRIA
6. Coron JM (1956) Control and nonlinearity. American Mathematical Society, Providence
7. Friedrich BM, Riedel-Kruse IH, Howard J, Jülicher F (2010) High-precision tracking of sperm swimming fine structure provides strong test of resistive force theory. J Exp Biol 213:1226–1234
8. Gebremedhin A, Pothen A, Walther A (eds) (2008) Exploiting sparsity in jacobian computation via coloring and automatic differentiation: a case study in a simulated moving bed process. In: Proceedings of the fifth international conference on automatic differentiation

9. Gray J, Hancock J (1955) The propulsion of sea-urchin spermatozoa. J Exp Biol 32:802–814
10. Happel J, Brenner H (1965) Low Reynolds number hydrodynamics with special applications to particulate media. Prentice-Hall, Englewood Cliffs, NJ
11. Jurdjevic V (1997) Geometric control theory. Cambridge University Press, Cambridge
12. Purcell E (1977) Life at low Reynolds number. Am J Phys 45:3–11
13. Riedel-Kruse IH, Hilfinger A, Howard J, Jülicher F (2007) How molecular motors shape the flagellar beat. HFSP J 1(3):192–208
14. Trelat E (2005) Contrôle optimal: théorie and applications. Collection Mathématiques Concrètes, Vuibert
15. Wächter A, Biegler L (2006) On the implementation of a primal-dual interior point filter line search algorithm for large-scale nonlinear programming. Math Program 106(1):25–57
16. Walther A, Griewank A (2012) Getting started with ADOL-C. Combinatorial scientific computing. Chapman-Hall/ CRC, London

Index

A

abstract Cauchy problem 56
activation 144, 148
    active strain 145
    loss of 152
    parameter of 149
anisotropy 17

B

Bessel potential space 56
biological system 93
BOCOP 177
boundary conditions 33
    Robin-type 32, 41
Brownian bridge 53

C

cancer 109
cell 14
    migration 74
    potency 74
    proliferation 74
    sperm 168
    stem 73
cell average 33
    intrinsic 33
    volumetric 33

chemotaxis 47, 74
collision
    generalized invariant 103
    invariant 103
    operator 101
complex network 94
controllability 171
curvature approximation 167

D

DLVO theory 118
drug delivery 30, 110
dynamical plastic structure 94

E

epithelial-mesenchymal transition 73
equation
    advection-diffusion 116
    advection-reaction-diffusion 54, 74
    diffusion 3
    Langevin 116
    partial differential 2
    random ordinary differential 51
    reaction-diffusion 47, 48, 53
    reaction-diffusion-taxis 47
    stochastic differential 51
    Stokes 116
extracellular matrix 73

F

FEniCS 154
fiber 94
   alignment 94
   cross-link 94
   interconnected 94

G

Gauss floor function 78
geometric control 171
global well-posedness 49, 50, 55, 56
go-or-grow dichotomy 49
go-or-grow-or-decay 55

H

haptotaxis 47, 74
homeostasis 93
homogenization 27, 28, 32
   asymptotic 2, 27, 28, 32
   cell problem 36
   effective coefficients 38
   method of multiple scales 27, 28, 32
   near-periodic microstructure 27, 28, 32, 41
   solvability condition 35, 36
   transformation of the derivative 32
   transforming the normal 33
   treatment of the macroscale variable within a cell 39
hydraulic conductivity 23
hydrodynamic scaling 106

I

incompressibility 142
INF patterns 62, 64, 65
integrins 75
interstitial gap 47

L

level set function 33

Lie algebra 173
limit
   asymptotic 94
   large scale 94
limiter
   monotonized central 80

M

macroscopic uniformity 18
mass conservation 95
membrane potential 53
mesoscopic description 94
method
   embedded multiscale 132
   finite volume $\sim$ with exponential fitting 121
   implicit-explicit Runge-Kutta 81
   Krylov subspace 81
   lattice kinetic Monte Carlo 121
   level set 33
mild solution 56
minimum time optimal control problem 176
model
   acid-mediated tumor invasion 45
   agent-based 95
   continuum 94
   deterministic 46
   Happel's sphere-in-cell 119
   individual-based 94
   Keller-Segel 74
   Lotka-Volterra 54
   mean-field kinetic 98
   micro-macro 48, 49, 120
   micro-meso 48
   micro-meso-macro 48
   multiscale 48
   SAIG 51, 53–56, 59–65
   SMAMCI 51, 53, 55, 57, 58
   stochastic 46
   stochastic acid invasion with gaps 51
   stochastic multiscale acid-mediated cancer invasion 51
multiscale 1
muscle fibers 141

N

N-link swimmer 162
nematic alignment 95
Newtonian fluid 19

Index

O

Ornstein-Uhlenbeck process 53

P

passive energy 142
periodicity 2, 14
pH-taxis 47, 48, 51, 55
Piola-Kirchhoff stress tensor 143, 147
porous media 19
power series 6
Purcell stroke 178
Purcell's swimmer 177

R

Resistive Force Theory 163

S

sarcopenia 139, 152

scale
    macroscale 4, 126
    macroscopic 48
    mesoscale 48
    microscale 1, 4, 116
    microscopic 48
    subcellular 48
    time scale
        macroscopic 75
        microscopic 75
self-organization 93
structural tensor 142

T

time delay 76, 81
tissue
    drug delivery to 30
    transport in tumors 110
    transport theorem 41
    tumor heterogeneity 49

W

weak solution 55

Editorial Policy

1. Volumes in the following three categories will be published in LNCSE:

i) Research monographs
ii) Tutorials
iii) Conference proceedings

Those considering a book which might be suitable for the series are strongly advised to contact the publisher or the series editors at an early stage.

2. Categories i) and ii). Tutorials are lecture notes typically arising via summer schools or similar events, which are used to teach graduate students. These categories will be emphasized by Lecture Notes in Computational Science and Engineering. **Submissions by interdisciplinary teams of authors are encouraged.** The goal is to report new developments – quickly, informally, and in a way that will make them accessible to non-specialists. In the evaluation of submissions timeliness of the work is an important criterion. Texts should be well-rounded, well-written and reasonably self-contained. In most cases the work will contain results of others as well as those of the author(s). In each case the author(s) should provide sufficient motivation, examples, and applications. In this respect, Ph.D. theses will usually be deemed unsuitable for the Lecture Notes series. Proposals for volumes in these categories should be submitted either to one of the series editors or to Springer-Verlag, Heidelberg, and will be refereed. A provisional judgement on the acceptability of a project can be based on partial information about the work: a detailed outline describing the contents of each chapter, the estimated length, a bibliography, and one or two sample chapters – or a first draft. A final decision whether to accept will rest on an evaluation of the completed work which should include

– at least 100 pages of text;
– a table of contents;
– an informative introduction perhaps with some historical remarks which should be accessible to readers unfamiliar with the topic treated;
– a subject index.

3. Category iii). Conference proceedings will be considered for publication provided that they are both of exceptional interest and devoted to a single topic. One (or more) expert participants will act as the scientific editor(s) of the volume. They select the papers which are suitable for inclusion and have them individually refereed as for a journal. Papers not closely related to the central topic are to be excluded. Organizers should contact the Editor for CSE at Springer at the planning stage, see *Addresses* below.

In exceptional cases some other multi-author-volumes may be considered in this category.

4. Only works in English will be considered. For evaluation purposes, manuscripts may be submitted in print or electronic form, in the latter case, preferably as pdf- or zipped ps-files. Authors are requested to use the LaTeX style files available from Springer at http://www.springer.com/gp/authors-editors/book-authors-editors/manuscript-preparation/5636 (Click on LaTeX Template → monographs or contributed books).

For categories ii) and iii) we strongly recommend that all contributions in a volume be written in the same LaTeX version, preferably LaTeX2e. Electronic material can be included if appropriate. Please contact the publisher.

Careful preparation of the manuscripts will help keep production time short besides ensuring satisfactory appearance of the finished book in print and online.

5. The following terms and conditions hold. Categories i), ii) and iii):

Authors receive 50 free copies of their book. No royalty is paid.
Volume editors receive a total of 50 free copies of their volume to be shared with authors, but no royalties.

Authors and volume editors are entitled to a discount of 33.3 % on the price of Springer books purchased for their personal use, if ordering directly from Springer.

6. Springer secures the copyright for each volume.

Addresses:

Timothy J. Barth
NASA Ames Research Center
NAS Division
Moffett Field, CA 94035, USA
barth@nas.nasa.gov

Michael Griebel
Institut für Numerische Simulation
der Universität Bonn
Wegelerstr. 6
53115 Bonn, Germany
griebel@ins.uni-bonn.de

David E. Keyes
Mathematical and Computer Sciences
and Engineering
King Abdullah University of Science
and Technology
P.O. Box 55455
Jeddah 21534, Saudi Arabia
david.keyes@kaust.edu.sa

and

Department of Applied Physics
and Applied Mathematics
Columbia University
500 W. 120 th Street
New York, NY 10027, USA
kd2112@columbia.edu

Risto M. Nieminen
Department of Applied Physics
Aalto University School of Science
and Technology
00076 Aalto, Finland
risto.nieminen@aalto.fi

Dirk Roose
Department of Computer Science
Katholieke Universiteit Leuven
Celestijnenlaan 200A
3001 Leuven-Heverlee, Belgium
dirk.roose@cs.kuleuven.be

Tamar Schlick
Department of Chemistry
and Courant Institute
of Mathematical Sciences
New York University
251 Mercer Street
New York, NY 10012, USA
schlick@nyu.edu

Editor for Computational Science
and Engineering at Springer:
Martin Peters
Springer-Verlag
Mathematics Editorial IV
Tiergartenstrasse 17
69121 Heidelberg, Germany
martin.peters@springer.com

Lecture Notes in Computational Science and Engineering

1. D. Funaro, *Spectral Elements for Transport-Dominated Equations*.
2. H.P. Langtangen, *Computational Partial Differential Equations*. Numerical Methods and Diffpack Programming.
3. W. Hackbusch, G. Wittum (eds.), *Multigrid Methods V*.
4. P. Deuflhard, J. Hermans, B. Leimkuhler, A.E. Mark, S. Reich, R.D. Skeel (eds.), *Computational Molecular Dynamics: Challenges, Methods, Ideas*.
5. D. Kröner, M. Ohlberger, C. Rohde (eds.), *An Introduction to Recent Developments in Theory and Numerics for Conservation Laws*.
6. S. Turek, *Efficient Solvers for Incompressible Flow Problems*. An Algorithmic and Computational Approach.
7. R. von Schwerin, ***Mu**lti **B**ody **S**ystem **SIM**ulation*. Numerical Methods, Algorithms, and Software.
8. H.-J. Bungartz, F. Durst, C. Zenger (eds.), *High Performance Scientific and Engineering Computing*.
9. T.J. Barth, H. Deconinck (eds.), *High-Order Methods for Computational Physics*.
10. H.P. Langtangen, A.M. Bruaset, E. Quak (eds.), *Advances in Software Tools for Scientific Computing*.
11. B. Cockburn, G.E. Karniadakis, C.-W. Shu (eds.), *Discontinuous Galerkin Methods*. Theory, Computation and Applications.
12. U. van Rienen, *Numerical Methods in Computational Electrodynamics*. Linear Systems in Practical Applications.
13. B. Engquist, L. Johnsson, M. Hammill, F. Short (eds.), *Simulation and Visualization on the Grid*.
14. E. Dick, K. Riemslagh, J. Vierendeels (eds.), *Multigrid Methods VI*.
15. A. Frommer, T. Lippert, B. Medeke, K. Schilling (eds.), *Numerical Challenges in Lattice Quantum Chromodynamics*.
16. J. Lang, *Adaptive Multilevel Solution of Nonlinear Parabolic PDE Systems*. Theory, Algorithm, and Applications.
17. B.I. Wohlmuth, *Discretization Methods and Iterative Solvers Based on Domain Decomposition*.
18. U. van Rienen, M. Günther, D. Hecht (eds.), *Scientific Computing in Electrical Engineering*.
19. I. Babuška, P.G. Ciarlet, T. Miyoshi (eds.), *Mathematical Modeling and Numerical Simulation in Continuum Mechanics*.
20. T.J. Barth, T. Chan, R. Haimes (eds.), *Multiscale and Multiresolution Methods*. Theory and Applications.
21. M. Breuer, F. Durst, C. Zenger (eds.), *High Performance Scientific and Engineering Computing*.
22. K. Urban, *Wavelets in Numerical Simulation*. Problem Adapted Construction and Applications.
23. L.F. Pavarino, A. Toselli (eds.), *Recent Developments in Domain Decomposition Methods*.

24. T. Schlick, H.H. Gan (eds.), *Computational Methods for Macromolecules: Challenges and Applications*.

25. T.J. Barth, H. Deconinck (eds.), *Error Estimation and Adaptive Discretization Methods in Computational Fluid Dynamics*.

26. M. Griebel, M.A. Schweitzer (eds.), *Meshfree Methods for Partial Differential Equations*.

27. S. Müller, *Adaptive Multiscale Schemes for Conservation Laws*.

28. C. Carstensen, S. Funken, W. Hackbusch, R.H.W. Hoppe, P. Monk (eds.), *Computational Electromagnetics*.

29. M.A. Schweitzer, *A Parallel Multilevel Partition of Unity Method for Elliptic Partial Differential Equations*.

30. T. Biegler, O. Ghattas, M. Heinkenschloss, B. van Bloemen Waanders (eds.), *Large-Scale PDE-Constrained Optimization*.

31. M. Ainsworth, P. Davies, D. Duncan, P. Martin, B. Rynne (eds.), *Topics in Computational Wave Propagation. Direct and Inverse Problems*.

32. H. Emmerich, B. Nestler, M. Schreckenberg (eds.), *Interface and Transport Dynamics*. Computational Modelling.

33. H.P. Langtangen, A. Tveito (eds.), *Advanced Topics in Computational Partial Differential Equations*. Numerical Methods and Diffpack Programming.

34. V. John, *Large Eddy Simulation of Turbulent Incompressible Flows*. Analytical and Numerical Results for a Class of LES Models.

35. E. Bänsch (ed.), *Challenges in Scientific Computing - CISC 2002*.

36. B.N. Khoromskij, G. Wittum, *Numerical Solution of Elliptic Differential Equations by Reduction to the Interface*.

37. A. Iske, *Multiresolution Methods in Scattered Data Modelling*.

38. S.-I. Niculescu, K. Gu (eds.), *Advances in Time-Delay Systems*.

39. S. Attinger, P. Koumoutsakos (eds.), *Multiscale Modelling and Simulation*.

40. R. Kornhuber, R. Hoppe, J. Périaux, O. Pironneau, O. Wildlund, J. Xu (eds.), *Domain Decomposition Methods in Science and Engineering*.

41. T. Plewa, T. Linde, V.G. Weirs (eds.), *Adaptive Mesh Refinement – Theory and Applications*.

42. A. Schmidt, K.G. Siebert, *Design of Adaptive Finite Element Software. The Finite Element Toolbox ALBERTA*.

43. M. Griebel, M.A. Schweitzer (eds.), *Meshfree Methods for Partial Differential Equations II*.

44. B. Engquist, P. Lötstedt, O. Runborg (eds.), *Multiscale Methods in Science and Engineering*.

45. P. Benner, V. Mehrmann, D.C. Sorensen (eds.), *Dimension Reduction of Large-Scale Systems*.

46. D. Kressner, *Numerical Methods for General and Structured Eigenvalue Problems*.

47. A. Boriçi, A. Frommer, B. Joó, A. Kennedy, B. Pendleton (eds.), *QCD and Numerical Analysis III*.

48. F. Graziani (ed.), *Computational Methods in Transport*.

49. B. Leimkuhler, C. Chipot, R. Elber, A. Laaksonen, A. Mark, T. Schlick, C. Schütte, R. Skeel (eds.), *New Algorithms for Macromolecular Simulation*.

50. M. Bücker, G. Corliss, P. Hovland, U. Naumann, B. Norris (eds.), *Automatic Differentiation: Applications, Theory, and Implementations.*

51. A.M. Bruaset, A. Tveito (eds.), *Numerical Solution of Partial Differential Equations on Parallel Computers.*

52. K.H. Hoffmann, A. Meyer (eds.), *Parallel Algorithms and Cluster Computing.*

53. H.-J. Bungartz, M. Schäfer (eds.), *Fluid-Structure Interaction.*

54. J. Behrens, *Adaptive Atmospheric Modeling.*

55. O. Widlund, D. Keyes (eds.), *Domain Decomposition Methods in Science and Engineering XVI.*

56. S. Kassinos, C. Langer, G. Iaccarino, P. Moin (eds.), *Complex Effects in Large Eddy Simulations.*

57. M. Griebel, M.A Schweitzer (eds.), *Meshfree Methods for Partial Differential Equations III.*

58. A.N. Gorban, B. Kégl, D.C. Wunsch, A. Zinovyev (eds.), *Principal Manifolds for Data Visualization and Dimension Reduction.*

59. H. Ammari (ed.), *Modeling and Computations in Electromagnetics: A Volume Dedicated to Jean-Claude Nédélec.*

60. U. Langer, M. Discacciati, D. Keyes, O. Widlund, W. Zulehner (eds.), *Domain Decomposition Methods in Science and Engineering XVII.*

61. T. Mathew, *Domain Decomposition Methods for the Numerical Solution of Partial Differential Equations.*

62. F. Graziani (ed.), *Computational Methods in Transport: Verification and Validation.*

63. M. Bebendorf, *Hierarchical Matrices. A Means to Efficiently Solve Elliptic Boundary Value Problems.*

64. C.H. Bischof, H.M. Bücker, P. Hovland, U. Naumann, J. Utke (eds.), *Advances in Automatic Differentiation.*

65. M. Griebel, M.A. Schweitzer (eds.), *Meshfree Methods for Partial Differential Equations IV.*

66. B. Engquist, P. Lötstedt, O. Runborg (eds.), *Multiscale Modeling and Simulation in Science.*

67. I.H. Tuncer, Ü. Gülcat, D.R. Emerson, K. Matsuno (eds.), *Parallel Computational Fluid Dynamics 2007.*

68. S. Yip, T. Diaz de la Rubia (eds.), *Scientific Modeling and Simulations.*

69. A. Hegarty, N. Kopteva, E. O'Riordan, M. Stynes (eds.), *BAIL 2008 – Boundary and Interior Layers.*

70. M. Bercovier, M.J. Gander, R. Kornhuber, O. Widlund (eds.), *Domain Decomposition Methods in Science and Engineering XVIII.*

71. B. Koren, C. Vuik (eds.), *Advanced Computational Methods in Science and Engineering.*

72. M. Peters (ed.), *Computational Fluid Dynamics for Sport Simulation.*

73. H.-J. Bungartz, M. Mehl, M. Schäfer (eds.), *Fluid Structure Interaction II - Modelling, Simulation, Optimization.*

74. D. Tromeur-Dervout, G. Brenner, D.R. Emerson, J. Erhel (eds.), *Parallel Computational Fluid Dynamics 2008.*

75. A.N. Gorban, D. Roose (eds.), *Coping with Complexity: Model Reduction and Data Analysis.*

76. J.S. Hesthaven, E.M. Rønquist (eds.), *Spectral and High Order Methods for Partial Differential Equations.*

77. M. Holtz, *Sparse Grid Quadrature in High Dimensions with Applications in Finance and Insurance.*

78. Y. Huang, R. Kornhuber, O.Widlund, J. Xu (eds.), *Domain Decomposition Methods in Science and Engineering XIX.*

79. M. Griebel, M.A. Schweitzer (eds.), *Meshfree Methods for Partial Differential Equations V.*

80. P.H. Lauritzen, C. Jablonowski, M.A. Taylor, R.D. Nair (eds.), *Numerical Techniques for Global Atmospheric Models.*

81. C. Clavero, J.L. Gracia, F.J. Lisbona (eds.), *BAIL 2010 – Boundary and Interior Layers, Computational and Asymptotic Methods.*

82. B. Engquist, O. Runborg, Y.R. Tsai (eds.), *Numerical Analysis and Multiscale Computations.*

83. I.G. Graham, T.Y. Hou, O. Lakkis, R. Scheichl (eds.), *Numerical Analysis of Multiscale Problems.*

84. A. Logg, K.-A. Mardal, G. Wells (eds.), *Automated Solution of Differential Equations by the Finite Element Method.*

85. J. Blowey, M. Jensen (eds.), *Frontiers in Numerical Analysis - Durham 2010.*

86. O. Kolditz, U.-J. Gorke, H. Shao, W. Wang (eds.), *Thermo-Hydro-Mechanical-Chemical Processes in Fractured Porous Media - Benchmarks and Examples.*

87. S. Forth, P. Hovland, E. Phipps, J. Utke, A. Walther (eds.), *Recent Advances in Algorithmic Differentiation.*

88. J. Garcke, M. Griebel (eds.), *Sparse Grids and Applications.*

89. M. Griebel, M.A. Schweitzer (eds.), *Meshfree Methods for Partial Differential Equations VI.*

90. C. Pechstein, *Finite and Boundary Element Tearing and Interconnecting Solvers for Multiscale Problems.*

91. R. Bank, M. Holst, O. Widlund, J. Xu (eds.), *Domain Decomposition Methods in Science and Engineering XX.*

92. H. Bijl, D. Lucor, S. Mishra, C. Schwab (eds.), *Uncertainty Quantification in Computational Fluid Dynamics.*

93. M. Bader, H.-J. Bungartz, T. Weinzierl (eds.), *Advanced Computing.*

94. M. Ehrhardt, T. Koprucki (eds.), *Advanced Mathematical Models and Numerical Techniques for Multi-Band Effective Mass Approximations.*

95. M. Azaïez, H. El Fekih, J.S. Hesthaven (eds.), *Spectral and High Order Methods for Partial Differential Equations ICOSAHOM 2012.*

96. F. Graziani, M.P. Desjarlais, R. Redmer, S.B. Trickey (eds.), *Frontiers and Challenges in Warm Dense Matter.*

97. J. Garcke, D. Pflüger (eds.), *Sparse Grids and Applications – Munich 2012.*

98. J. Erhel, M. Gander, L. Halpern, G. Pichot, T. Sassi, O. Widlund (eds.), *Domain Decomposition Methods in Science and Engineering XXI.*

99. R. Abgrall, H. Beaugendre, P.M. Congedo, C. Dobrzynski, V. Perrier, M. Ricchiuto (eds.), *High Order Nonlinear Numerical Methods for Evolutionary PDEs - HONOM 2013.*

100. M. Griebel, M.A. Schweitzer (eds.), *Meshfree Methods for Partial Differential Equations VII.*

101. R. Hoppe (ed.), *Optimization with PDE Constraints - OPTPDE 2014*.

102. S. Dahlke, W. Dahmen, M. Griebel, W. Hackbusch, K. Ritter, R. Schneider, C. Schwab, H. Yserentant (eds.), *Extraction of Quantifiable Information from Complex Systems*.

103. A. Abdulle, S. Deparis, D. Kressner, F. Nobile, M. Picasso (eds.), *Numerical Mathematics and Advanced Applications - ENUMATH 2013*.

104. T. Dickopf, M.J. Gander, L. Halpern, R. Krause, L.F. Pavarino (eds.), *Domain Decomposition Methods in Science and Engineering XXII*.

105. M. Mehl, M. Bischoff, M. Schäfer (eds.), *Recent Trends in Computational Engineering - CE2014*. Optimization, Uncertainty, Parallel Algorithms, Coupled and Complex Problems.

106. R.M. Kirby, M. Berzins, J.S. Hesthaven (eds.), *Spectral and High Order Methods for Partial Differential Equations - ICOSAHOM'14*.

107. B. Jüttler, B. Simeon (eds.), *Isogeometric Analysis and Applications 2014*.

108. P. Knobloch (ed.), *Boundary and Interior Layers, Computational and Asymptotic Methods – BAIL 2014*.

109. J. Garcke, D. Pflüger (eds.), *Sparse Grids and Applications – Stuttgart 2014*.

110. H. P. Langtangen, *Finite Difference Computing with Exponential Decay Models*.

111. A. Tveito, G.T. Lines, *Computing Characterizations of Drugs for Ion Channels and Receptors Using Markov Models*.

112. B. Karazösen, M. Manguoğlu, M. Tezer-Sezgin, S. Göktepe, Ö. Uğur (eds.), *Numerical Mathematics and Advanced Applications - ENUMATH 2015*.

113. H.-J. Bungartz, P. Neumann, W.E. Nagel (eds.), *Software for Exascale Computing - SPPEXA 2013-2015*.

114. G.R. Barrenechea, F. Brezzi, A. Cangiani, E.H. Georgoulis (eds.), *Building Bridges: Connections and Challenges in Modern Approaches to Numerical Partial Differential Equations*.

115. M. Griebel, M.A. Schweitzer (eds.), *Meshfree Methods for Partial Differential Equations VIII*.

116. C.-O. Lee, X.-C. Cai, D.E. Keyes, H.H. Kim, A. Klawonn, E.-J. Park, O.B. Widlund (eds.), *Domain Decomposition Methods in Science and Engineering XXIII*.

117. T. Sakurai, S.-L. Zhang, T. Imamura, Y. Yamamoto, Y. Kuramashi, T. Hoshi (eds.), *Eigenvalue Problems: Algorithms, Software and Applications in Petascale Computing*. EPASA 2015, Tsukuba, Japan, September 2015.

118. T. Richter (ed.), *Fluid-structure Interactions*. Models, Analysis and Finite Elements.

119. M.L. Bittencourt, N.A. Dumont, J.S. Hesthaven (eds.), *Spectral and High Order Methods for Partial Differential Equations ICOSAHOM 2016*. Selected Papers from the ICOSAHOM Conference, June 27-July 1, 2016, Rio de Janeiro, Brazil.

120. Z. Huang, M. Stynes, Z. Zhang (eds.), *Boundary and Interior Layers, Computational and Asymptotic Methods BAIL 2016*.

121. S.P.A. Bordas, E.N. Burman, M.G. Larson, M.A. Olshanskii (eds.), *Geometrically Unfitted Finite Element Methods and Applications*. Proceedings of the UCL Workshop 2016.

122. A. Gerisch, R. Penta, J. Lang (eds.), *Multiscale Models in Mechano and Tumor Biology*. Modeling, Homogenization, and Applications.

For further information on these books please have a look at our mathematics catalogue at the following URL: www.springer.com/series/3527

Monographs in Computational Science and Engineering

1. J. Sundnes, G.T. Lines, X. Cai, B.F. Nielsen, K.-A. Mardal, A. Tveito, *Computing the Electrical Activity in the Heart*.

For further information on this book, please have a look at our mathematics catalogue at the following URL: www.springer.com/series/7417

Texts in Computational Science and Engineering

1. H. P. Langtangen, *Computational Partial Differential Equations. Numerical Methods and Diffpack Programming*. 2nd Edition

2. A. Quarteroni, F. Saleri, P. Gervasio, *Scientific Computing with MATLAB and Octave*. 4th Edition

3. H. P. Langtangen, *Python Scripting for Computational Science*. 3rd Edition

4. H. Gardner, G. Manduchi, *Design Patterns for e-Science*.

5. M. Griebel, S. Knapek, G. Zumbusch, *Numerical Simulation in Molecular Dynamics*.

6. H. P. Langtangen, *A Primer on Scientific Programming with Python*. 5th Edition

7. A. Tveito, H. P. Langtangen, B. F. Nielsen, X. Cai, *Elements of Scientific Computing*.

8. B. Gustafsson, *Fundamentals of Scientific Computing*.

9. M. Bader, *Space-Filling Curves*.

10. M. Larson, F. Bengzon, *The Finite Element Method: Theory, Implementation and Applications*.

11. W. Gander, M. Gander, F. Kwok, *Scientific Computing: An Introduction using Maple and MATLAB*.

12. P. Deuflhard, S. Röblitz, *A Guide to Numerical Modelling in Systems Biology*.

13. M. H. Holmes, *Introduction to Scientific Computing and Data Analysis*.

14. S. Linge, H. P. Langtangen, *Programming for Computations - A Gentle Introduction to Numerical Simulations with MATLAB/Octave*.

15. S. Linge, H. P. Langtangen, *Programming for Computations - A Gentle Introduction to Numerical Simulations with Python*.

16. H.P. Langtangen, S. Linge, *Finite Difference Computing with PDEs - A Modern Software Approach*.

17. B. Gustafsson, *Hyponormal Quantization of Planar Domains*.

18. J. A. Trangenstein, *Scientific Computing*. Volume I - Linear and Nonlinear Equations.

19. J. A. Trangenstein, *Scientific Computing*. Volume II - Eigenvalues and Optimization.
20. J. A. Trangenstein, *Scientific Computing*. Volume III - Approximation and Integration.

For further information on these books please have a look at our mathematics catalogue at the following URL: www.springer.com/series/5151